IMAGES
of America

WASHINGTON'S SUNSET HIGHWAY

On the Cover: Denny Creek Camp was constructed by the US Forest Service in 1919. The side road from the Sunset Highway to the campground was at the foot of the long descent from Snoqualmie Pass (or long ascent up the pass, if one was heading east). Located at the turnoff, this store/gas station/garage/café served both highway travelers and local vacationers who chose the campground as a destination resort. (Courtesy of David Dilgard and Lisa Labovitch, Everett Public Library, J477.)

IMAGES
of America

WASHINGTON'S SUNSET HIGHWAY

Chuck Flood

ARCADIA
PUBLISHING

ISBN 978-1-4671-3203-9

Published by Arcadia Publishing
Charleston, South Carolina

Printed in the United States of America

Library of Congress Control Number: 2013957192

For all general information, please contact Arcadia Publishing:
Telephone 843-853-2070
Fax 843-853-0044
E-mail sales@arcadiapublishing.com
For customer service and orders:
Toll-Free 1-888-313-2665

Visit us on the Internet at www.arcadiapublishing.com

Another one for Debbie and the open road.

Contents

Acknowledgments

Credit is gratefully acknowledged to the following individuals and organizations for their assistance and enthusiastic support of this project:

Lori Luderman, Douglas County Historical Museum [DCHS]
Regina Tipton-Llamas, Ellensburg Public Library [EllensburgPL]
David Dilgard and Lisa Labovitch, Everett Public Library [EverettPL]
Erica Maniez, Issaquah Historical Society [IHS]
Jack O'Donnell [JCO'D]
Sadie Thayer, Kittitas County Historical Museum [KCHM]
Tannis Jeschke, Lincoln County Historical Museum [LCHM]
Rose Krause, Special Collections, Museum of Arts and Culture [MAC]
Ann Schempp, Moses Lake Museum & Art Center [MLM]
Carolyn Marr, Museum of History and Industry [MOHAI]
Elizabeth Stewart, Renton History Museum [RHM]
Kylie Fullmer and Karen Clinesmith, Ritzville Public Library [RitzvillePL]
Erin Krake, Roslyn Public Library [RoslynPL]
Julie Irick, Seattle Municipal Archives [SMA]
Riva Dean, Spokane Public Library [SPL]
Jayne Singleton and Chuck King, Spokane Valley Heritage Museum [SVHM]
Cristy Lake, Snoqualmie Valley Museum [SVM]
Nicolette Bromberg and Kris Kinsey, Special Collections, University of Washington [UW]
Dave Lundgren and Amy Larsen, Waterville Historic Hotel [WHH]
Washington Rural Heritage [WRH]
Washington State Digital Archives [WSDA]
Joy Werlink and Fred Poyner, Washington State Historical Society [WSHS]
Angela Battalio-Bunker, Wenatchee Valley Museum & Cultural Center [WVM]

Images not individually credited are from the author's personal collection. Images with credit lines are courtesy of the organization or individual whose abbreviation appears in the above list, which should be used to determine the contributor's full name.

Introduction

The Yellowstone Trail, the Arrowhead Trail, the National Old Roads Trail, the Victory Highway, the Dixie Highway, the Old Spanish Trail: At one time the nation was crisscrossed with highways bearing colorful and descriptive names like these.

Some were national-transcontinental, east-west, or border-to-border. The Lincoln Highway and the National Old Roads Trail were probably the most famous east-west routes, while the Mississippi River Scenic Highway and the Lakes to Gulf Highway, among others, ran north-south. Some were regional, sometimes covering only a single state—the Missouri Cross-State Highway, for instance.

They all had their boosters—individuals, cities, and states advocating good roads, easy flow of commerce, and scenic attractions, always with an eye on collecting as many dollars as possible from travelers. Some routes had formal associations to push their interests, such as the Lincoln Highway Association and the National Old Roads Ocean-to-Ocean Highway Association. Others relied on local initiative.

Washington State hosted several of these "named" highways: the Pacific Highway, the great north-south transportation corridor west of the mountains; the Yellowstone Trail, from Spokane to Seattle via Walla Walla and Yakima; the National Parks Highway, which took a more direct route between Spokane and Seattle via Wenatchee; and the Sunset Highway, which in places followed the same route as the Yellowstone and National Parks Highways.

The ballyhooing and promoting could not obscure the fact that many of these early highways were simply labels applied to already-existing roads. Relatively few brand-new roads were constructed. The emphasis was on identifying and designating enough segments of road as possible to create a continuous thread of highway of several hundreds or thousands of miles. What construction was done was accomplished by local effort and funding, often piecemeal and consisting of maintaining and improving existing roadway.

The result of all these competing highway-building efforts was confusion and chaos. The same physical road was laid claim to by different routes, causing problems for travelers trying to figure out which route to take from point A to point B, particularly if points A and B were at opposite ends of the country. Since cities often contributed to construction efforts, they were inclined to thwart any attempts to reroute highways away from them. Uniform construction and safety standards were lacking; a highway might have good stretches of concrete pavement within and adjacent to cities and towns, but between towns the roads often reverted to muddy trails or cow-paths. Consistent standards regarding right-of-way were a problem—particularly out west where, if a road passed through a farm or ranch, the owner sometimes felt no compunction against building a fence (possibly without a gate) right across the route.

Things changed in the 1920s. Passage of the Federal Aid Highway Act in 1921 heavily involved the national government in highway planning and funding. Standards for highway construction were put in place. In 1925 the Bureau of Public Roads, predecessor to today's Federal Highway

Administration, approved a plan by the American Association of State Highway Officials to create a numbering system for national highways. Within a few years, highway names were replaced by numbers, which—if less romantic—were at least consistent.

Washington's early highway history was typical of other states. Though a spider's web of county roads connected cities and towns, up until 1900, railroads and rivers were the main transportation conduits. As settlement pushed farther away from natural travel routes, demand for additional roads began to increase. Most often the roads radiated out from centers of commerce to smaller towns, rather than between the larger towns and cities. Long-distance travel and freight shipment were not important factors (that is what railroads were for), and as for quality, the new roads were the same as the old: dusty in summer, muddy bogs in winter. For decades the old roads sufficed, particularly since the majority of people rarely traveled more than 20 miles from home.

There was another impediment: Washington is geographically divided by the Cascade Mountains. A water-level route through the Cascades existed via the Columbia River along the southern edge of the state, but that route did not benefit Seattle-to-Spokane travel. The roundabout way between those two cities using the Columbia River added hundreds of miles to the journey. As a result, the two sides of the state developed independent transportation networks with only a handful of cross-mountain passages.

Even the earliest freighters were looking for a more direct route across the Cascades. The key came with the discovery of Snoqualmie Pass in 1853. At just over 3,000 feet, the pass is the lowest mountain crossing in Washington. Within a few years of its discovery, a wagon road was forged across the pass to connect the Puget Sound region to the drier interior. The wagon road was improved over time, and by the dawn of the automobile age it was a practical if difficult route, connecting on either side of the pass with pre-existing roads.

In 1913, the Washington State Highway Board designated the Sunset Highway as a Primary State Road. Its route was described thus: "A highway starting from the Pacific Highway at Renton, Washington; thence over the most feasible route by the way of Snoqualmie Pass into the Yakima River Valley; thence by way of Wenatchee, over the most feasible route, through Waterville and Spokane, to the state boundary." The reason for the name Sunset is not clear; eastbound travelers could have called it the Sunrise Highway.

By the late 1920s, with the demise of highway names, the route carried the US 10 shield—a segment of a transcontinental highway between Detroit and Seattle. A few decades later, Highway 10 was realigned to a more direct path between Spokane and Seattle via a crossing of the Columbia River at Vantage (details about specific routes are found in individual chapter introductions).

As the highways became heavily traveled, entrepreneurs rose to the opportunities. Roadside restaurants and cafés, tourist parks, auto courts and motels, and gas stations sprang up. Particularly east of the mountains, tourist-oriented services tended to be clustered in and around towns, with plenty of open space between. Unfortunately, only a fraction of the roadside businesses pictured in this book survive in their original form; many have disappeared.

Today the only officially designated US 10 is a 565-mile stretch in Michigan and Wisconsin. In Washington, Interstate 90 has largely replaced the 1950s route of Highway 10. In many places, the interstate was built directly on top of the old highway.

But much remains to be explored. Ritzville has placed Historic Highway 10 markers along the route through town. The early alignment—the Sunset Highway through Waterville and Wenatchee—survives as Highway 2. Highway 97 follows much the same route as the Sunset Highway north from Cle Elum, and the old road over Blewett Pass offers today's adventurer a 1930s travel experience.

One

Idaho Border to Spokane

The journey starts at the western edge of Idaho, where the highway variously known as the Appleway, the Sunset Highway, and Highway 10 crosses the Spokane River into Washington State. The highway does not appear out of nowhere, of course; it is merely the westernmost extension of a roadway that originates in Michigan.

In pioneer days, the Spokane River was an active transportation corridor. Several roads such as the Mullan Road (built in the 1860s) converged on the river; a series of ferries and bridges conveyed homesteaders, freighters, miners, and soldiers on their trips between seacoast, mountains, and prairies.

The first bridge to carry motor traffic across the river at this point, built in the late 1890s, was closed soon after the auto age began and was not rebuilt until a dozen years later. Its replacement was lacking in capacity and safety to handle the rapid increase in traffic and in turn was replaced in 1939 by a bridge 1,200 feet upstream.

Once across the river, the highway traversed rich agricultural lands en route to Spokane passing through the communities of Greenacres, Veradale, Opportunity, and Dishman. The Appleway became Sprague Avenue at Greenacres; the highway stayed on Sprague Avenue all the way to Spokane. The 1939 travelers heading west found tourist courts and restaurants lining the highway in—and immediately adjacent to—these towns, but plenty of open space between.

Today, Interstate 90 has replaced the Appleway and Sprague Avenue for through traffic. Urban growth and development have spread eastward from Spokane to Greenacres; virtually all the farmlands are gone, and but few vestiges remain of the roadside businesses that catered to early travelers.

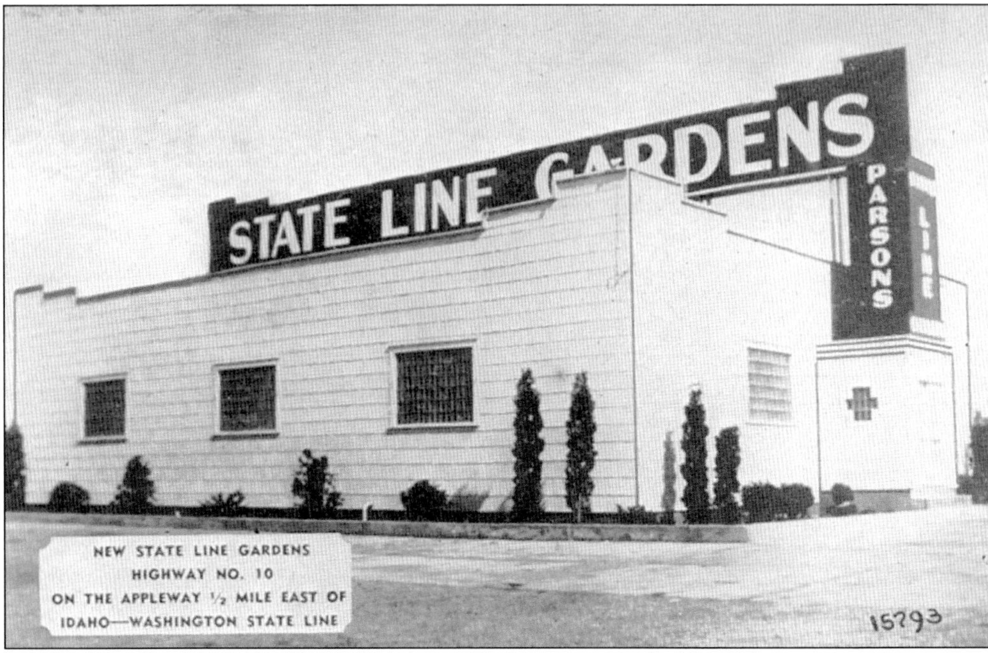

State Line Gardens was located just east of the Washington-Idaho border, where early liquor laws were somewhat lax. Father-and-son team J.R. and Preston Parsons operated the First-and-Last Tavern in 1935 but sold it and opened the Gardens prior to 1940. The building shown here was destroyed by fire on May 20, 1953. A new structure was opened the following July 1—it too is now gone.

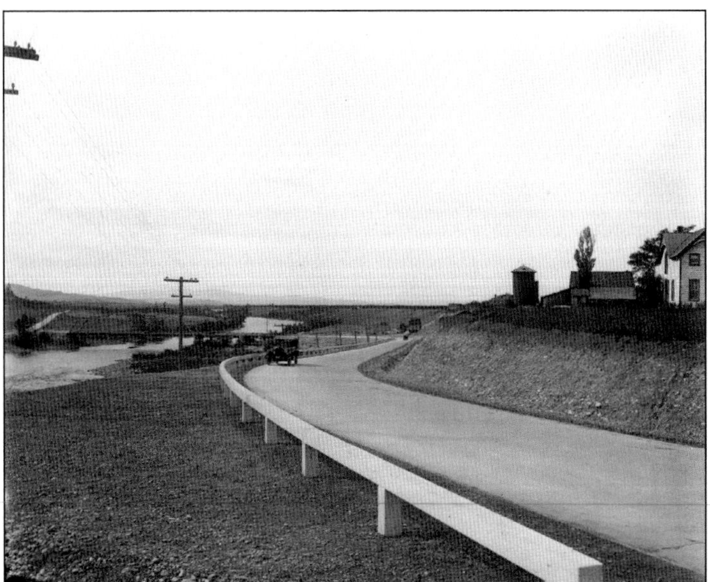

Bridges have spanned the Spokane River at the state border since 1864. In early days, a settlement called Spokane Bridge existed at the old wagon crossing. A truss bridge (seen here looking west) was constructed in 1911, replacing a bridge that failed after only six years of service, but as traffic increased, it too was deemed unsuitable due to dangerous curves where the highway approached the river crossing. (MAC, L95-111.21.)

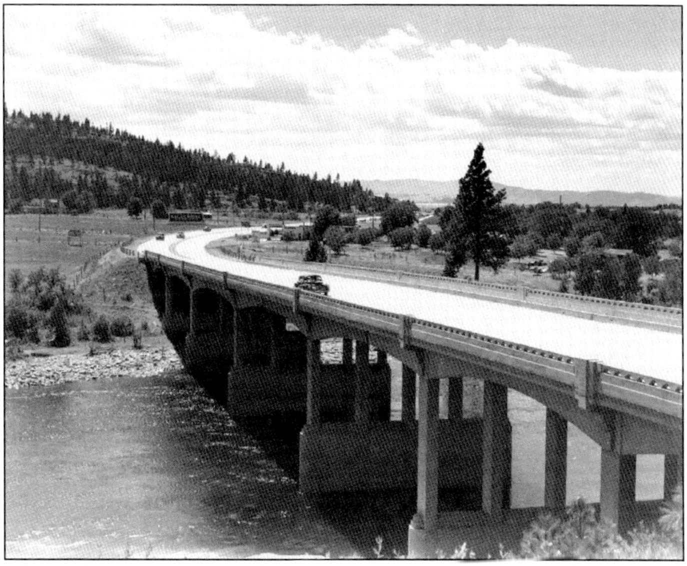

A replacement, the Appleway Bridge, was constructed in 1939 (the Appleway is the local name for the highway between Coeur d'Alene and Spokane). The reinforced-concrete structure was located 1,200 feet upstream from the old bridge. Its Art Deco design gave it a streamlined look. This view west towards Spokane shows a billboard advertising the Galax Hotel and a Chevron service station at the curve of the four-lane highway. (WSDA, AR-28001001-ph000175.)

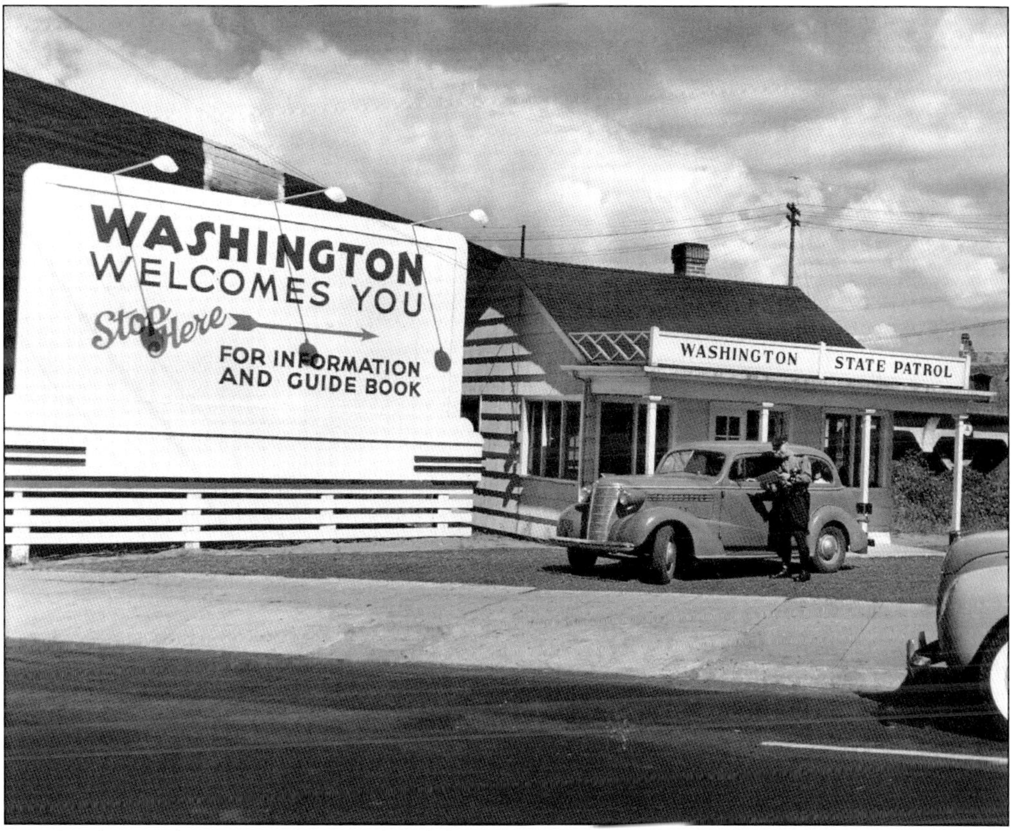

In 1937, the Washington State Patrol and the State Progress Commission opened welcome lodges along three major transcontinental highways as they entered the state at Spokane, Walla Walla, and Vancouver. The lodges were staffed with specially trained officers to provide tourists with information and maps. A common sight at state borders today, these information kiosks were a significant innovation at the time. (WSDA, AR-07809001-ph004872.)

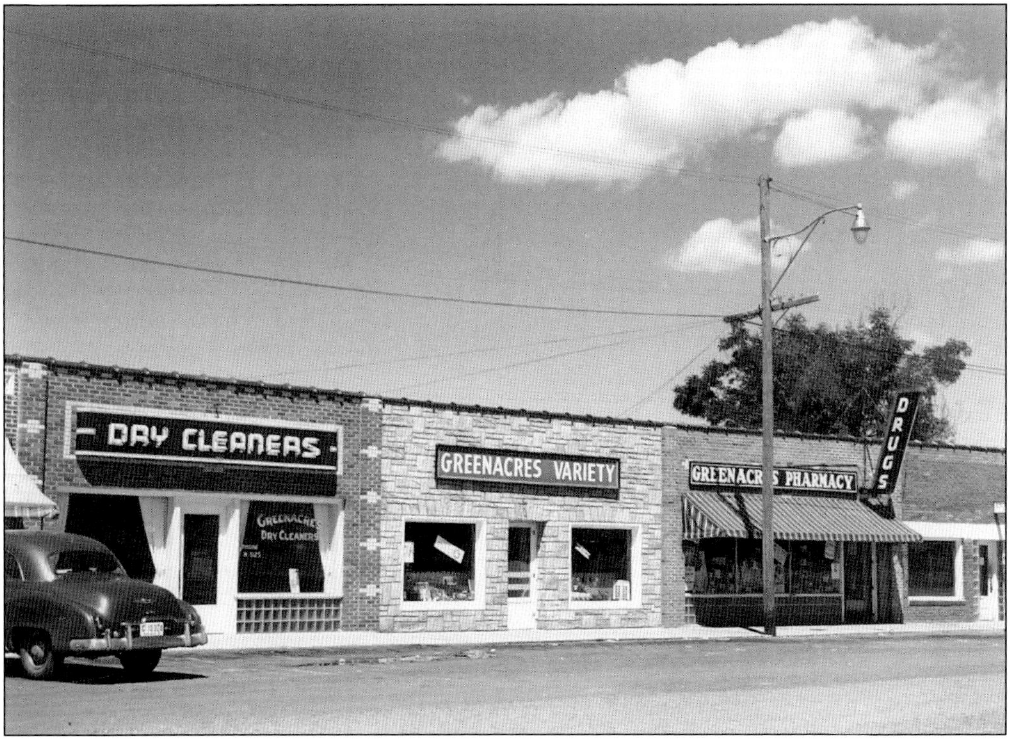

Greenacres, seven miles from the state border, was the first town westbound travelers encountered. Described in 1939 as "a community of garden tracts" with a compact business section, the town sat amid vast stretches of irrigated farmland, the highway lined with stands selling locally grown fruit and vegetables. In 2003, Greenacres joined with neighboring Opportunity and Veradale to incorporate as the City of Spokane Valley.

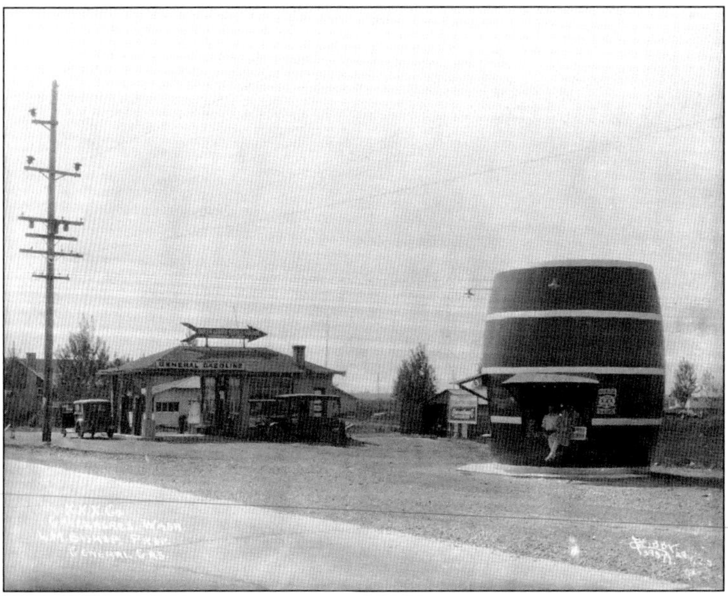

This Triple XXX root beer stand at Greenacres was one of many to be found along the early Sunset Highway. The distinctive barrel design made them landmarks for travelers, though this 1929 scene does not show much business—the only customer was a woman swinging her feet from her perch on the stand's counter. Adjacent was L.M. Bishop's Green Arrow Service Station and tourist camp. (MAC, L87-1.39337-29.)

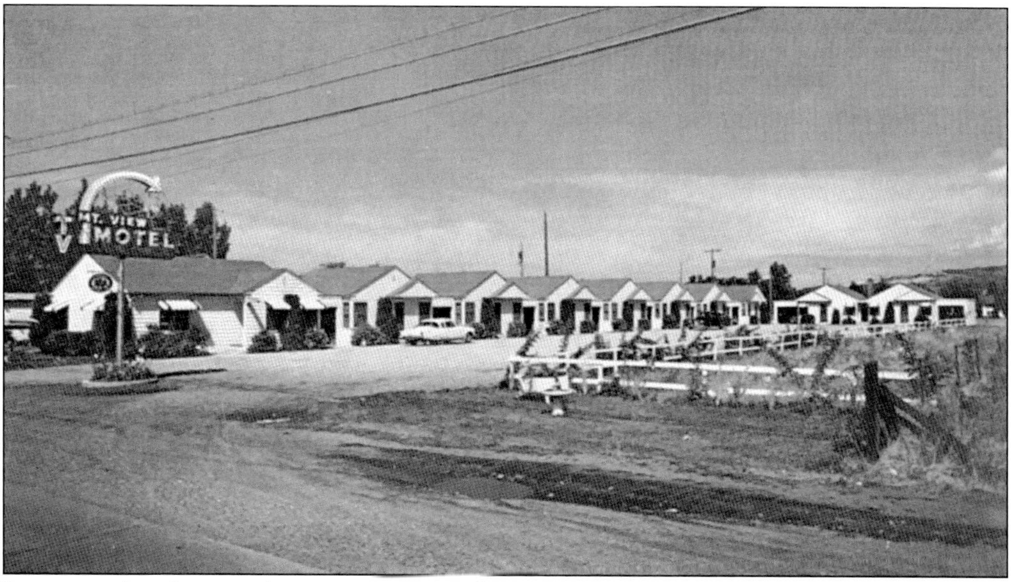

Promising cleanliness and courteous service, the Mountain View Motel was managed by Jeraldine Falconer in 1953. By the next year, ownership had passed to Keith and Henrietta McDonald and, in 1955, to Mr. and Mrs. Albert Neilson. The motel offered four-room units with kitchenettes, television, and garages. The AAA-approved motel was at 17111 E Sprague in Greenacres. A manufactured-homes sales lot occupies the site today. (SVHM, 2005.615.020.)

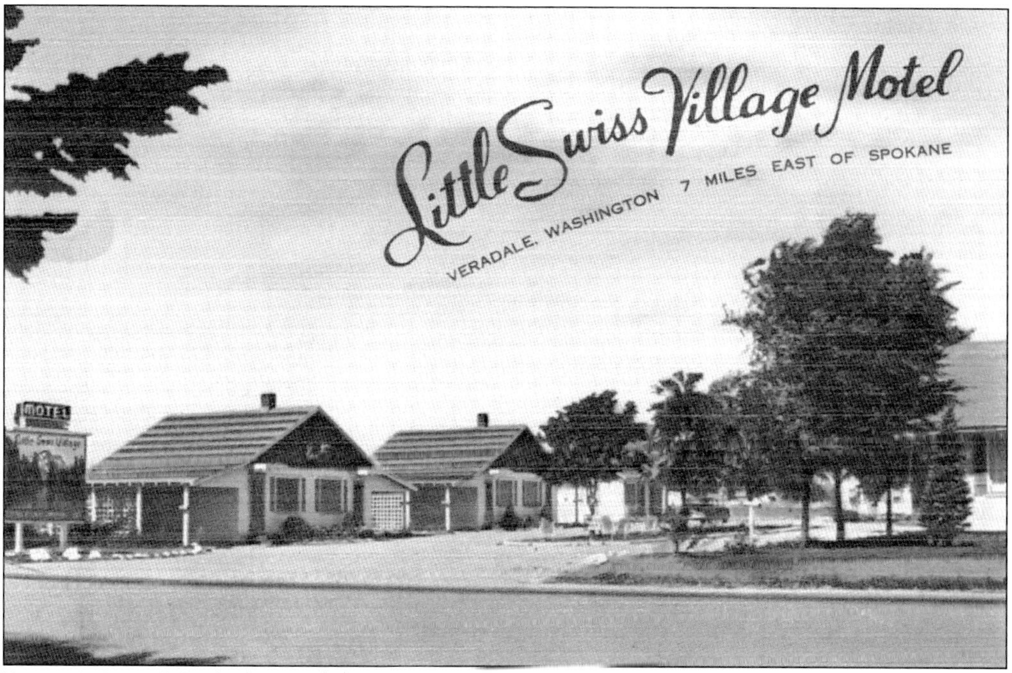

From 1949 to 1969, Arthur and Anna Gerdes owned and operated the Little Swiss Village Motel at 15609 E Sprague Avenue in Veradale, on Highway 10 seven miles east of Spokane. Colorfully striped roofs and Alpine styling enhanced the Swiss theme. The motel offered attractive modern units for two or more persons with sleeping rooms and electric kitchens. The site is now a business park.

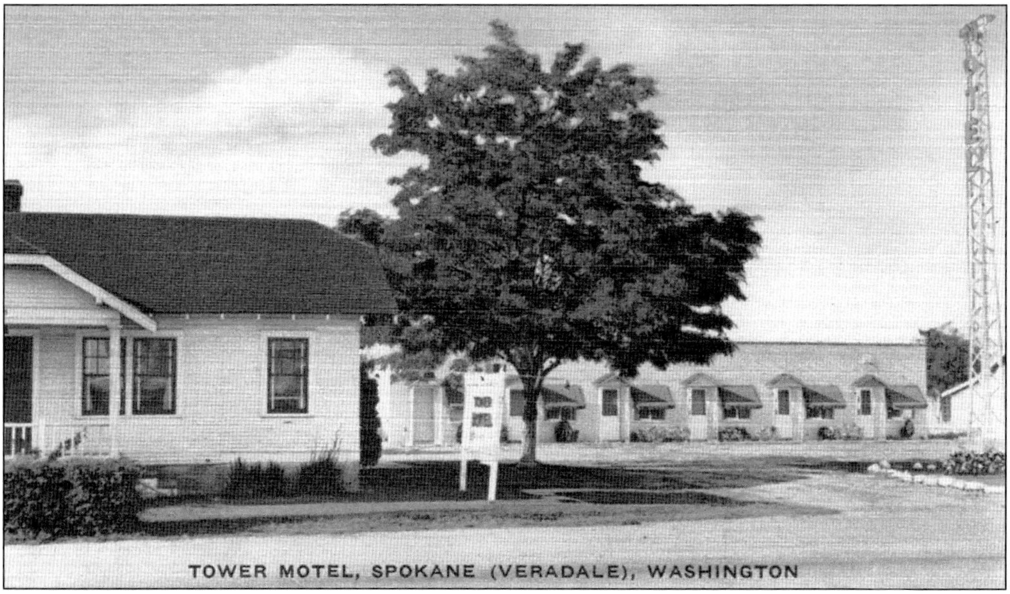

Mr. and Mrs. R.C. Arthur constructed a one-story motel—a single building of individual units—behind their white, red-roofed bungalow home. The units (singles or doubles) were fireproof and soundproof; each included an electric kitchen and a brightly-colored awning over the front window. A tower along the highway advertised the motel. The Tower was located at 15515 E Sprague in Opportunity. (SVHM, 2005.615.030.)

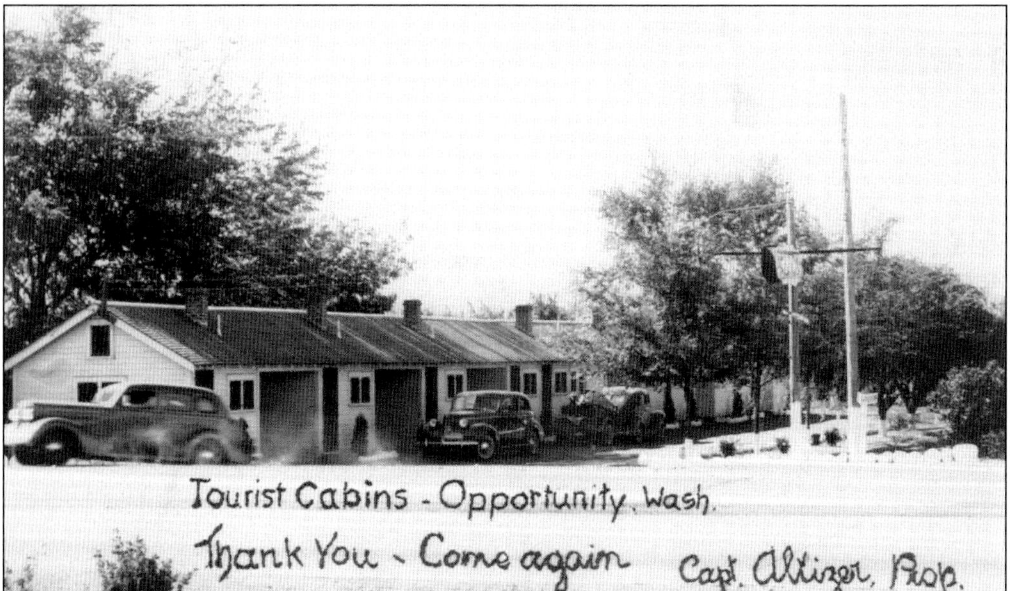

Captain Altizer's Opportunity Tourist Cabins, at 12902 E Sprague Avenue, consisted of a single row of eight units. Each unit had a toilet and shower, hot and cold running water, and an open garage. Nightly rates were $1–$1.35 for two people, $2–$3.20 for four. Listed in the 1941 Shell Directory of Auto Courts, the cabins are missing from the 1948 edition. (SVHM, 2005.615.029.)

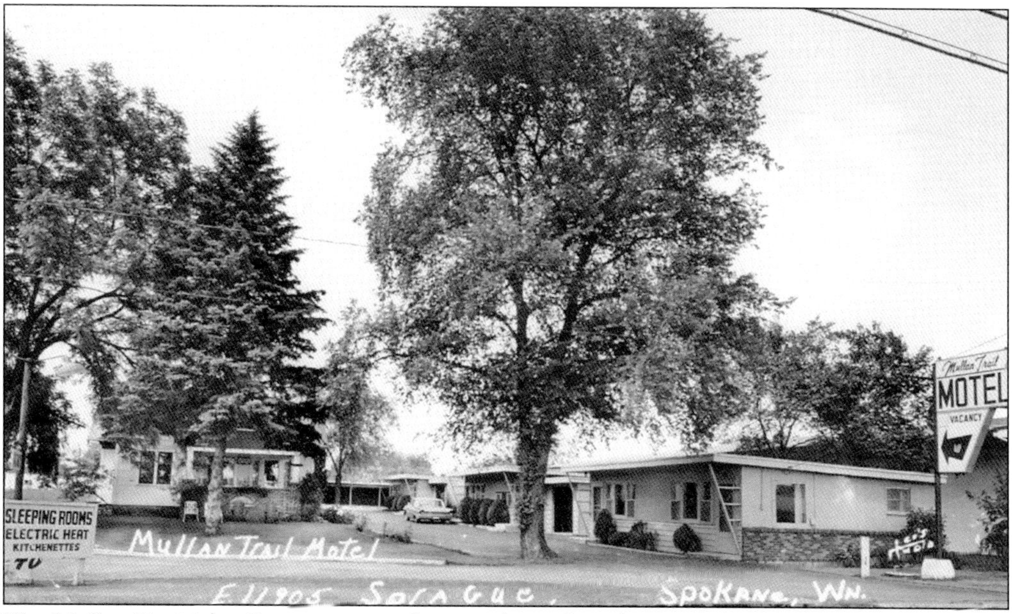

The Mullan Trail Motel's name pays tribute to the early military road between Fort Walla Walla, Washington, and Fort Benton, Montana. The road was constructed in 1859–1860 by Capt. John Mullan and, in the Spokane area, was more heavily used by freighters, miners, and homesteaders than soldiers. The motel, at 11905 E Sprague Avenue, was along a section of highway built on top of the old wagon road. (SVHM, 2005.615.021.)

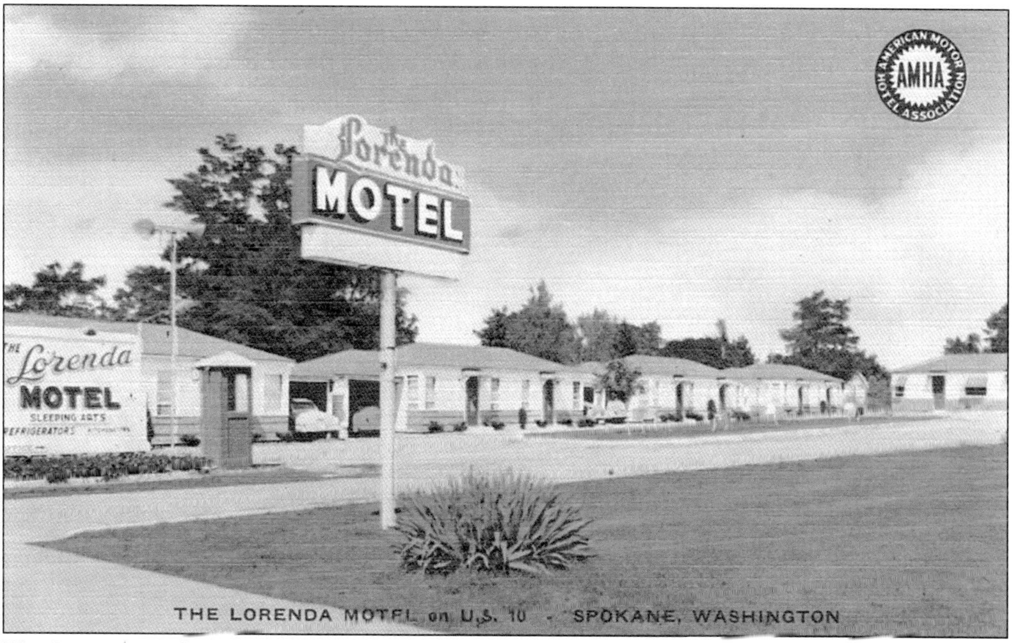

The Lorenda took its name from one of the owners, Charles and Lorena Aarli. Double cabins in a U-shaped configuration, with individual garages for each unit, surrounded an open green space at 11515 E Sprague, a 15-minute drive from downtown Spokane. The Aarlis sold the Lorenda to David and Diana Baker for approximately $80,000 in 1972. The new owners planned an expansion, but the motel fell victim to urban growth.

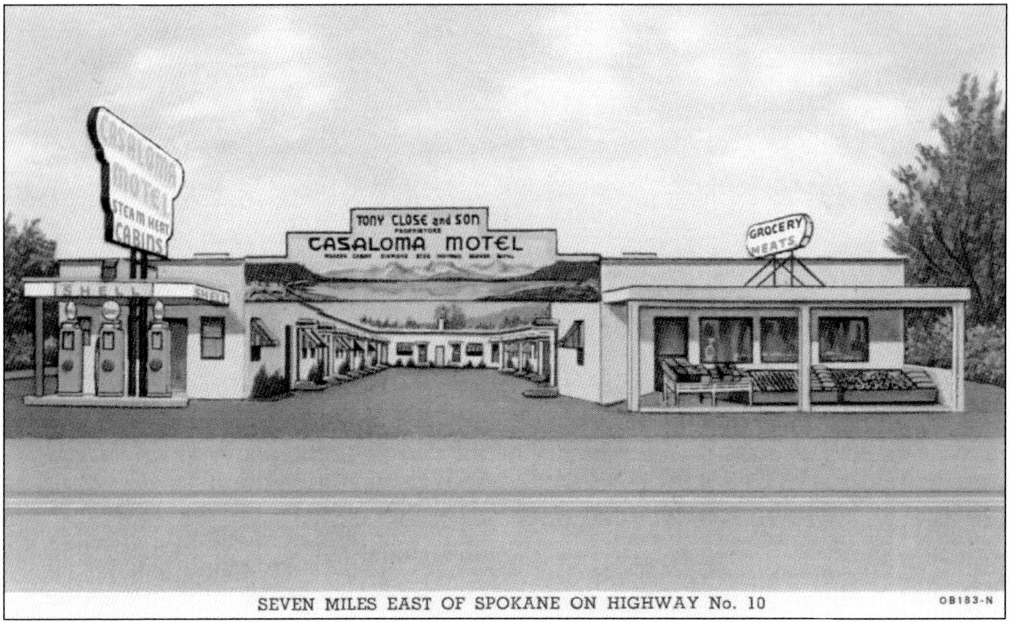

The Casaloma Motel contained 20 "motorists' apartments" housed in a horseshoe-shaped building set on three green acres. Each unit offered either three or five rooms and a modern kitchen. A Shell station and store occupied opposite ends of the horseshoe. Built in 1939 at a cost of $43,000, owner Tony Close contemplated additional construction the following year. The motel was located at 11124 E Sprague Avenue in Opportunity. (SVHM, 2007.615.300.)

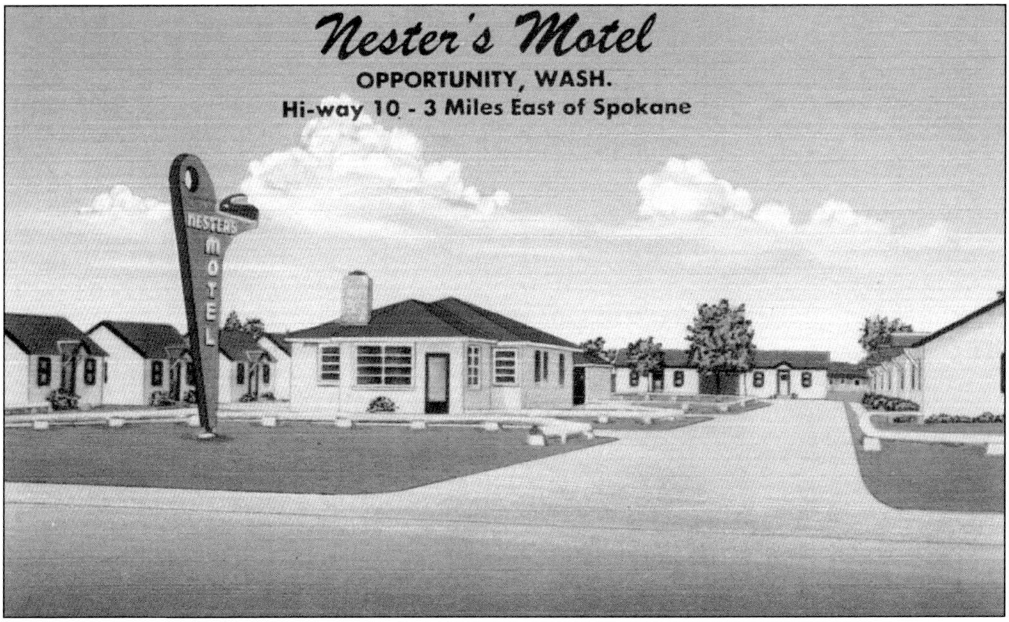

Nester's Motel, in business in 1949 under the ownership of Arthur M. and Irene Payne, advertised "sensible rates" and "a warm welcome to all tourist and commercial people." Located at 10915 E Sprague in Opportunity, Nester's consisted of 18 detached cabins set around a central office. In 1957, the Paynes traded the motel to Mr. and Mrs. Woodrow Wilson in exchange for a 2,300-acre ranch valued at $213,000. (SVHM, 2005.615.022.)

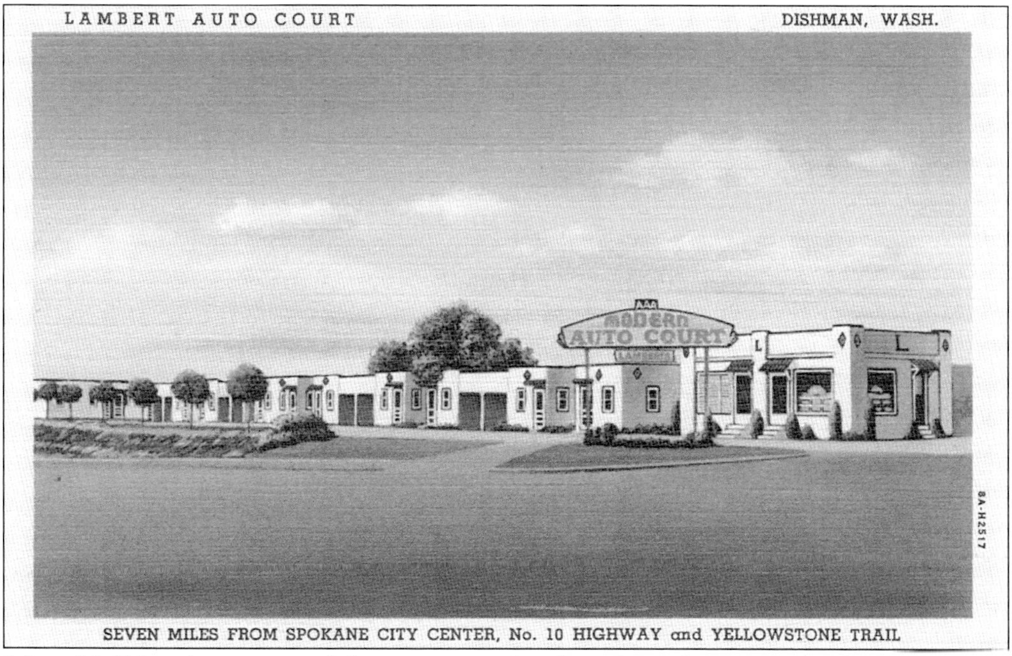

L.J. and L.F. Lambert managed the eponymous Lambert Auto Court outside of Dishman. The court offered 12 "strictly modern" one-, two-, and three-room units with tubs and showers, electric ranges, telephones, and private garages, at rates from $2 to $5 per night. The style of the court dates to the 1930s when usage of the Yellowstone Trail as a highway designation was no longer common.

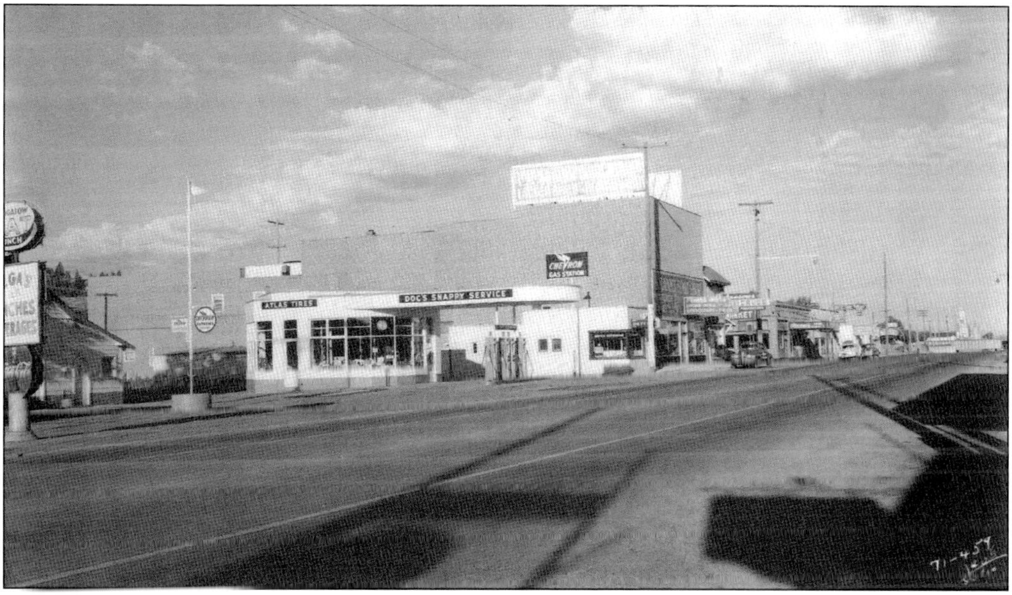

Thirteen miles west of the state border, the Appleway passed through Dishman, named for entrepreneur Addison T. Dishman and originally called Six Mile, which was its distance from Spokane. Though Dishman has become a suburb of the big city, farms once lined the Appleway almost to Spokane's city limits. This view of Dishman's business center in the 1940s shows the Bungalow Lunch and Doc's Snappy Service waiting for customers. (SPL.)

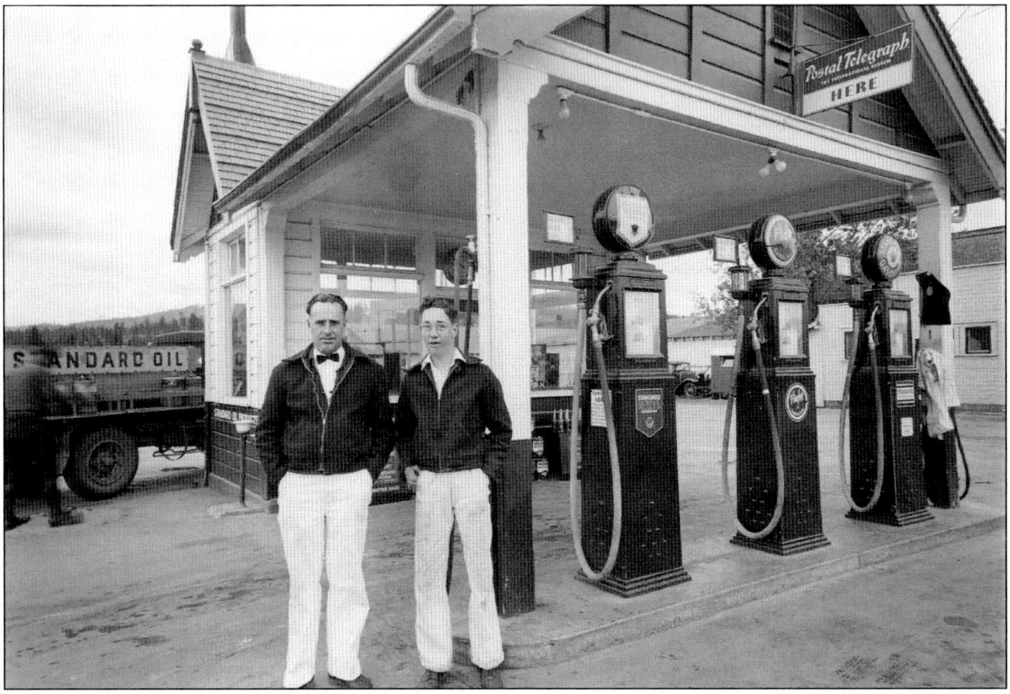

An earlier view of Doc's Snappy Service shows owner Harold C. Dockendorf and assistant Charles "Sparky" Burton ready to dispense Standard and Flight gasoline from their shiny, glass-topped pumps. Doc's was representative of a time when service stations attempted to blend in with their residential surroundings; the station is almost cottage-like in its architecture, with strong wood columns and decorative brackets in its portico. (SPL.)

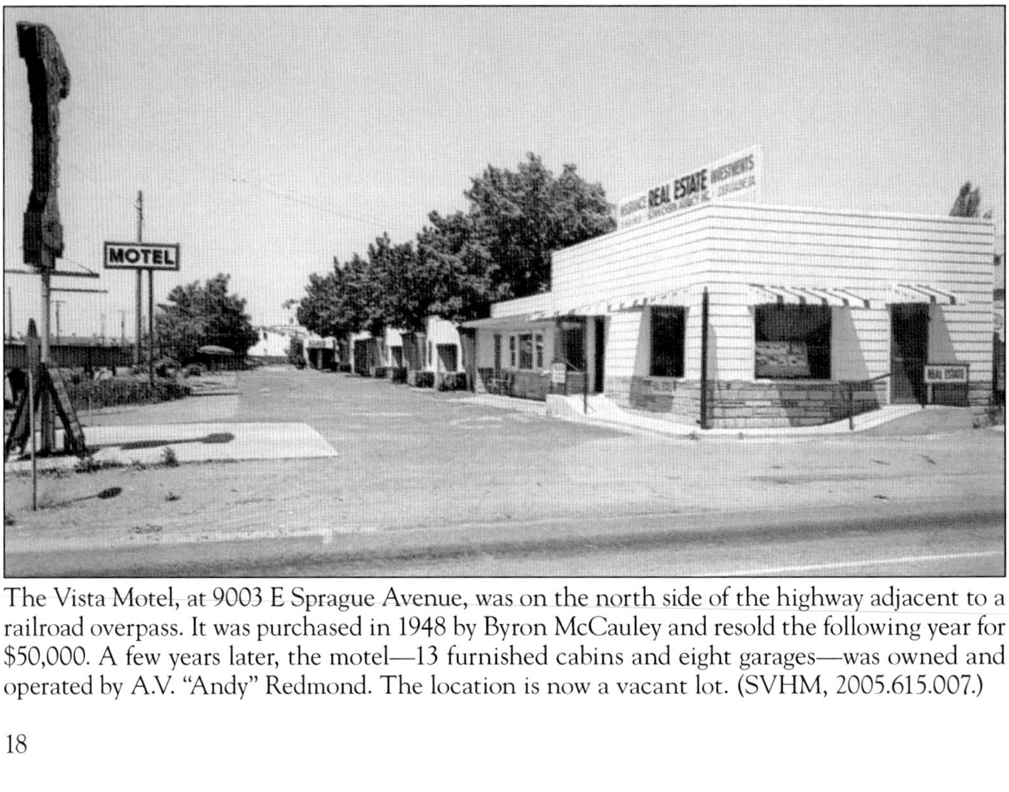

The Vista Motel, at 9003 E Sprague Avenue, was on the north side of the highway adjacent to a railroad overpass. It was purchased in 1948 by Byron McCauley and resold the following year for $50,000. A few years later, the motel—13 furnished cabins and eight garages—was owned and operated by A.V. "Andy" Redmond. The location is now a vacant lot. (SVHM, 2005.615.007.)

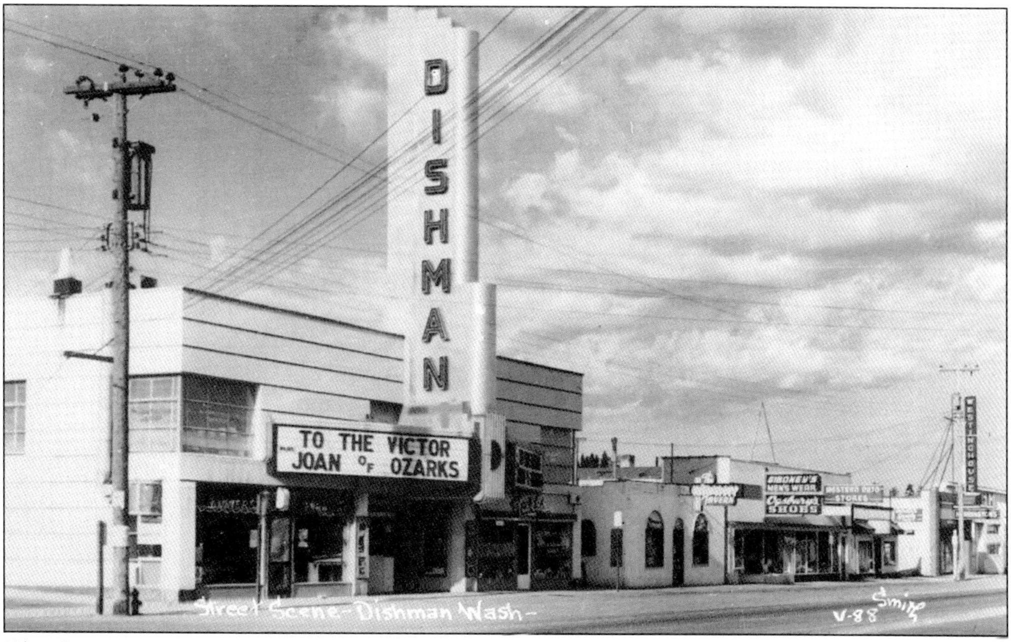

The 516-seat Dishman Theater celebrated its opening on November 6, 1938, with a showing of *King of the Turf* starring Dolores Costello and Adolphe Menjou. Built by town founder A.T. Dishman, the theater was operated by Mr. and Mrs. J. Collins. Clyde Dowell assumed management in 1943. A ticket for a cartoon, newsreel, and feature cost 35¢ in 1947. Today the theater, at 8803 E Sprague, is a gentlemen's club.

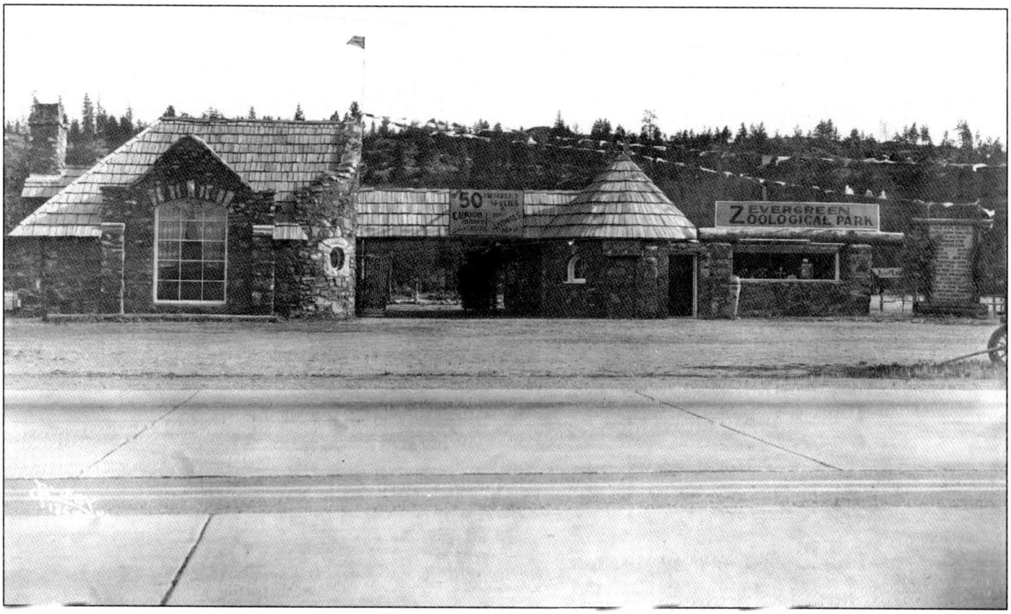

In 1939, A.E. Warsinske, deciding Spokane needed a zoo, opened his Evergreen Zoological Park at 8122 E Sprague. Built on 18 acres with a rustic entrance lodge, the zoo featured native animals and birds such as bears, coyotes, and cougars. The stone-fronted building was later converted into a restaurant, which bore half a dozen names over the years, being best known as the Plantation. (MAC, L87-1.18548-40.)

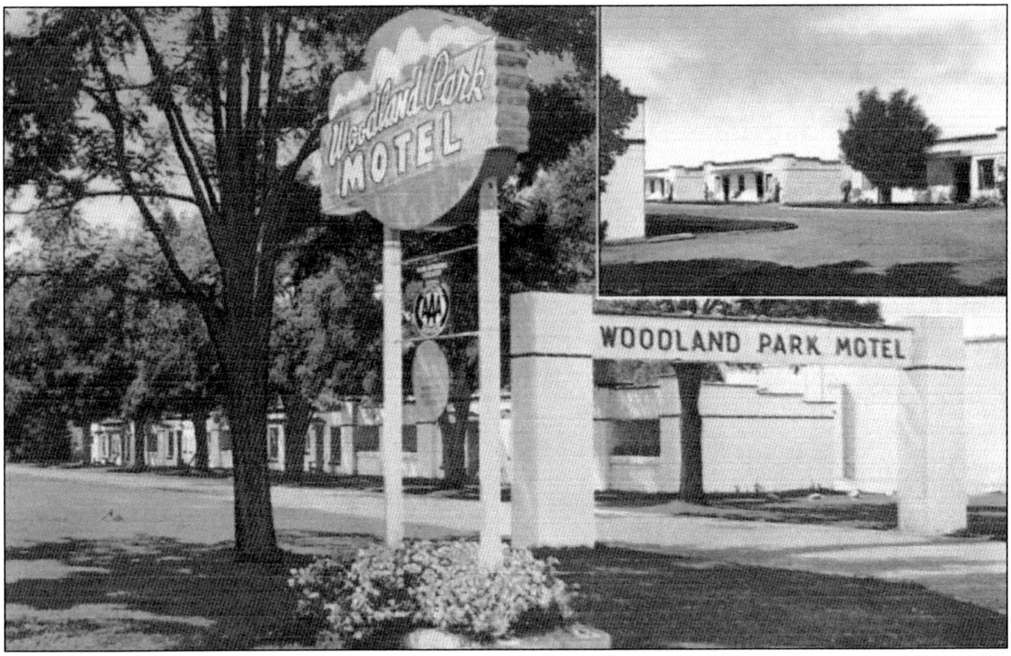

Shade trees and a green lawn offered a respite from Spokane's summer sun at the Woodland Park Motel. Located five-and-a-half miles east of city center at 8022 E Sprague Avenue, the entryway passed under a welcoming arch and along stucco-covered rows of apartments and standard motel rooms. The AAA-approved motel was managed for many years by Richard and Kathryn Lentz. Today, a car lot occupies the site.

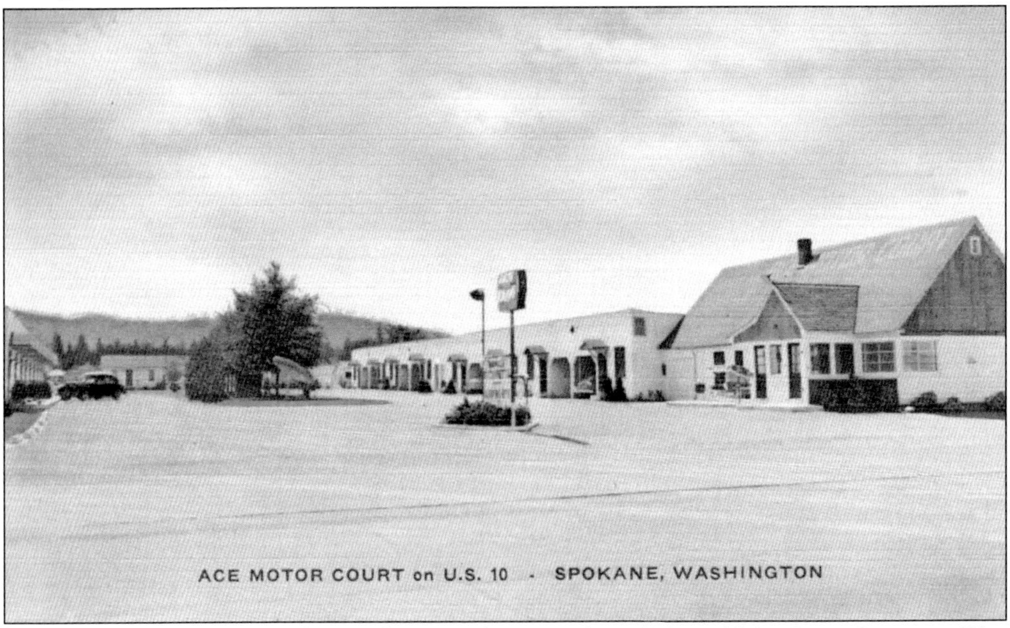

The Ace Motor Court opened around 1947 under ownership of Mr. and Mrs. Soren Sorenson. The Ace had eight single and eight double units with kitchenettes, tile baths, and air conditioning, plus a two-room residence for the managers. In 1949, the motel was sold to Lindsey and Frances Jackson. By 1957, the Ace had disappeared; the site, at 7210 E Sprague, is another used car lot.

Povey's Steak and Chicken Inn was an upscale dinner house at 6416 E Sprague. In this 1940 photograph, a jukebox gleams in the background; the wide floor invites dancing. By 1954, Povey's had moved to a new location seven miles east at E 15916 Sprague. Renamed Povey's Veradale Inn, it continued serving chicken and steak dinners. The old location became La Salle's Restaurant. (MAC, L87-1.20319-41.)

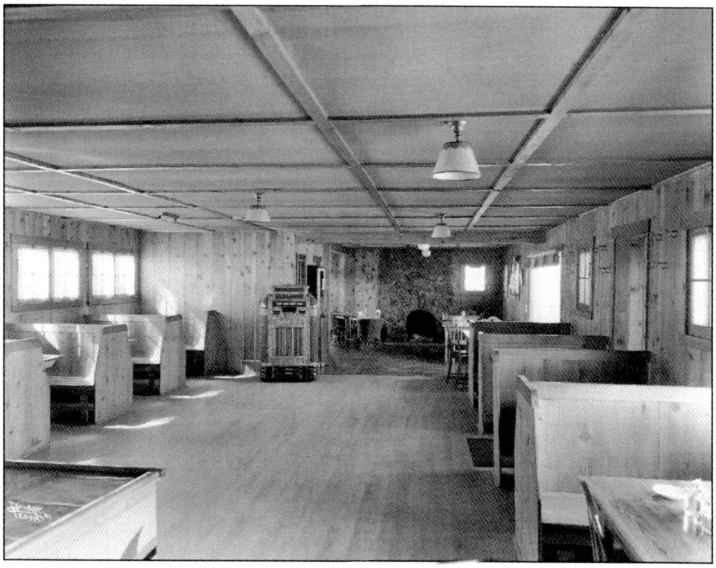

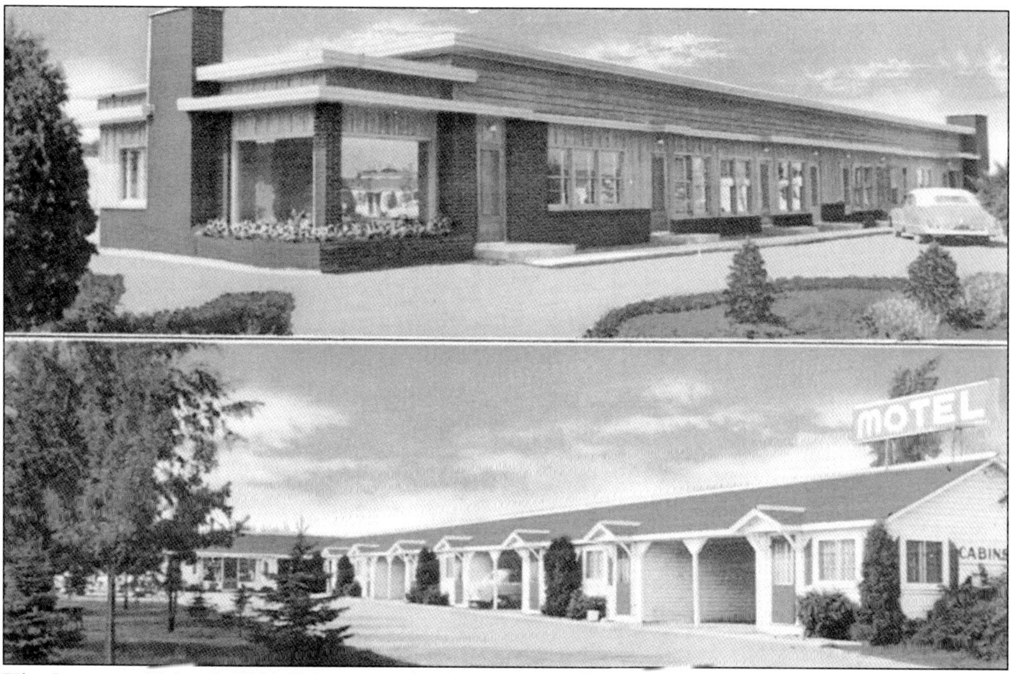

The Eastwood Motel, 5502 E Sprague, displayed two different styles of roadside architecture: early wood cabins separated by carports and a later brick building. Its 40 units were equipped with hot water and electric heat, televisions, and room phones. When the motel was destroyed by fire in February 1977, a guest saved his grand piano by covering it with a rubberized tarp while the building burned down around it.

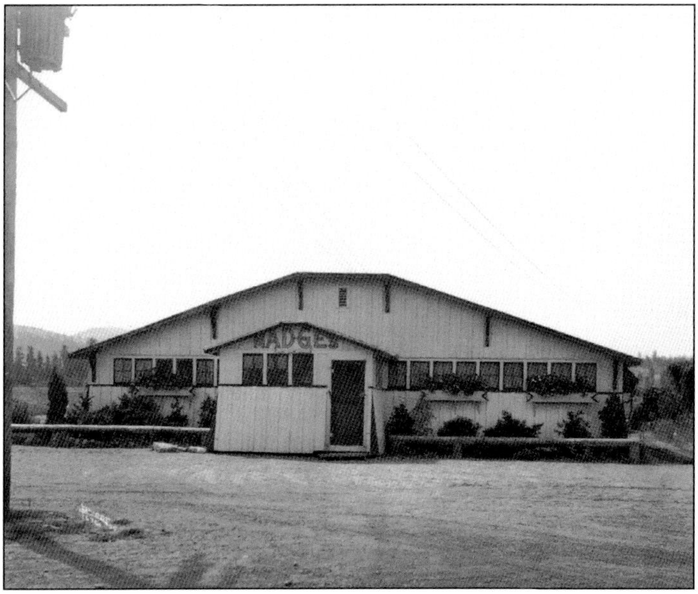

Madge and Guy Standish operated Madge's Chicken Dinner Inn until 1938, when an acrimonious divorce left Madge as sole owner. Located at 4928 E Sprague, Madge's was typical of the "dine and dance" roadhouses popping up around the country—large, commodious buildings usually located along highways on the outskirts of big cities. In 1945, when Madge's added dancing, the inn was under management of a Major Roberts. (MAC, L87-1.6751-35.)

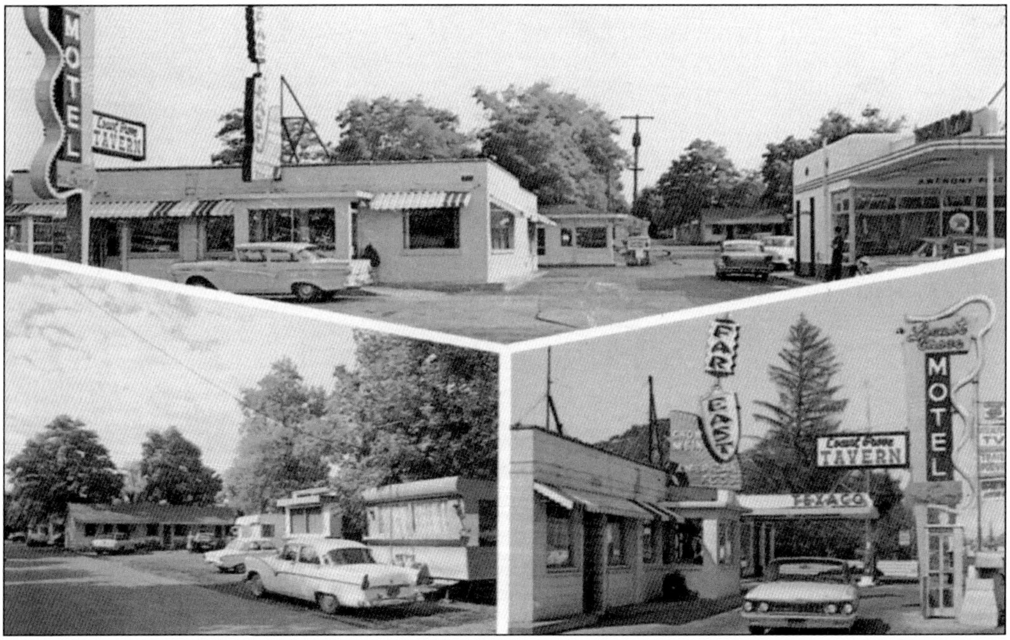

George Lareida and his father opened the Locust Grove Camp in 1917. By 1935, the camp consisted of 22 cottages at a nightly rate of \$1–\$2. By the late 1950s, the Lareidas had added a tavern, service station, and the Far East restaurant to the premises. The restaurant is still in business as the Dragon Garden; a few cabins still stand behind it at 4824 E Sprague Avenue. (SVHM, 2005.615.018.)

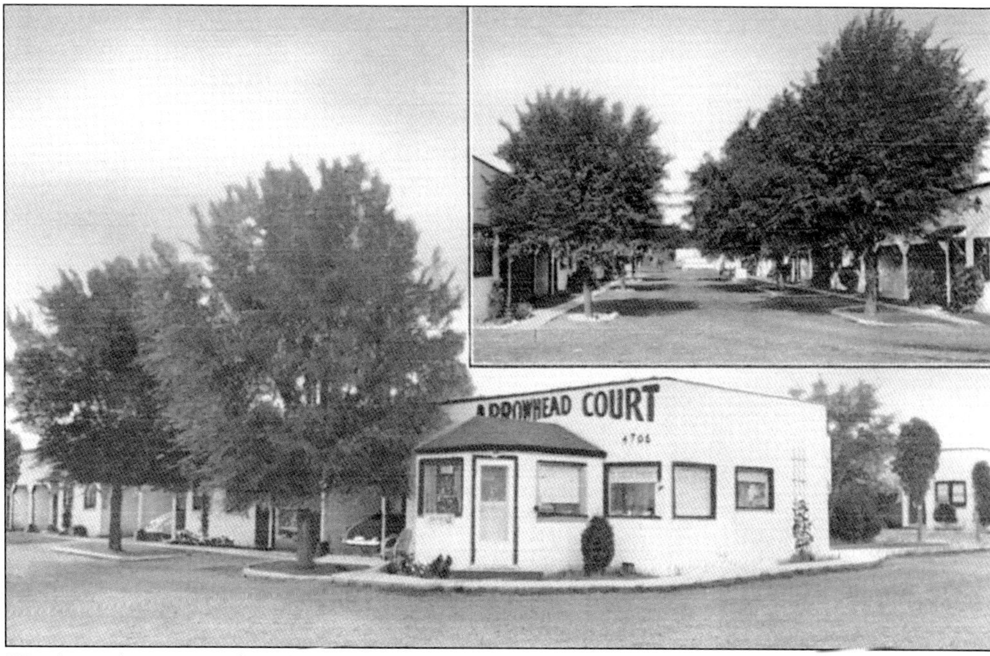

Westbound motorists found a cluster of motels just east of Spokane's city limits. First up was the Arrowhead Motor Court at 4706 E Sprague. Owned and managed by Dr. and Mrs. Charles A. Ham, the court consisted of 22 older-style sleeping cottages arranged in parallel tree-lined rows with intervening carports; each "ultra-modern housekeeping unit" came with a fully-equipped kitchenette. A check-cashing office is at this address today.

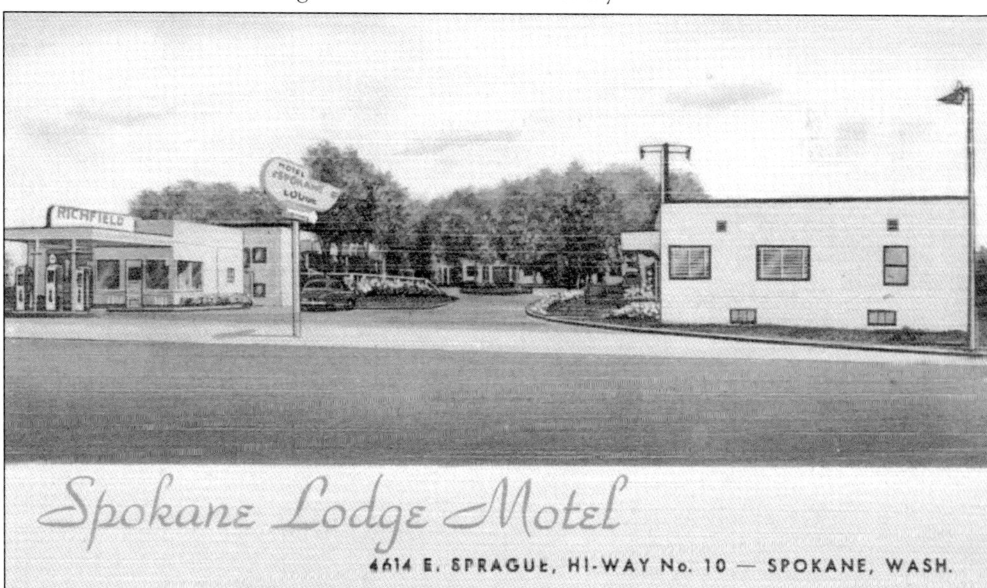

Next was the Spokane Lodge Motel at 4614 E Sprague Avenue, offering clean, modern, air-conditioned units, some with kitchenettes. The 19 units were arranged in a U-shape with a Richfield gas station sitting right at the motel's entrance. A member of the Washington Motel Association, in 1949, the Spokane Lodge was designated a "key court," one of several positioned at city entrances statewide to assist tourists in finding lodging.

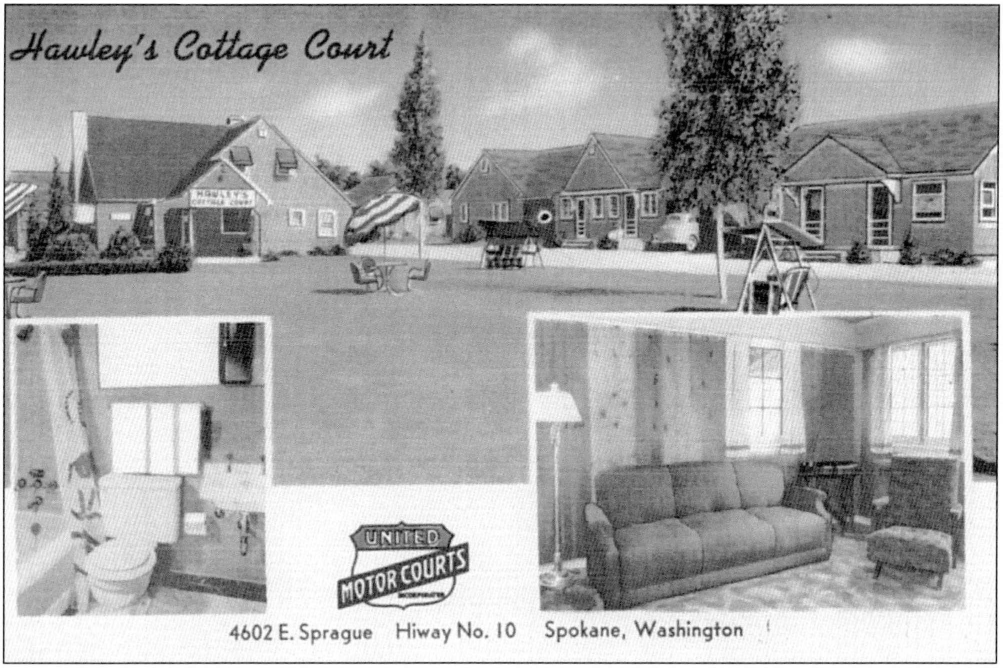

Hawley's Cottage Court featured 32 rooms and kitchenette apartments with overstuffed furniture, hot water heat, tubs and showers, room phones, and covered parking. Brightly colored lawn chairs and swings occupied an open green space between the two rows of units. Built by H.S. Hawley prior to 1941, the court was located at 4602 E Sprague Avenue. In *Lodging for a Night*, Duncan Hines recommended Hawley's as "first class in every way."

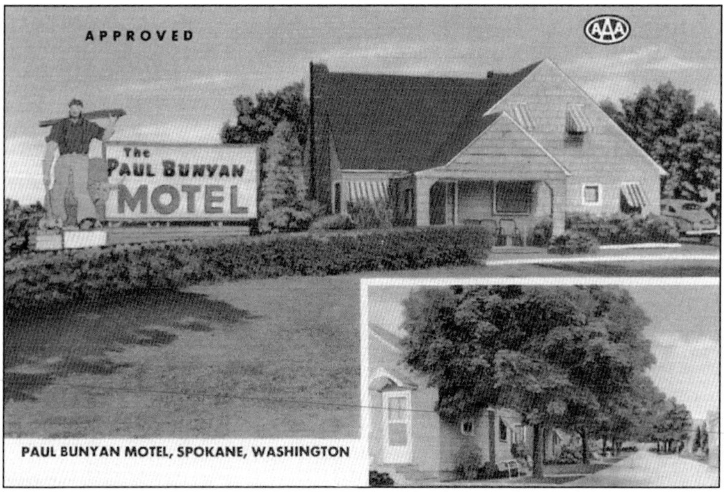

By 1980, Hawley's had been sold and renamed the Paul Bunyan Motel, complete with a 12-foot-tall cutout figure of the giant lumberjack himself. In that year, the motel figured in a double murder involving outlaw motorcycle gangs, stolen guns, and disappearing witnesses. The site of the motel is now a parking lot.

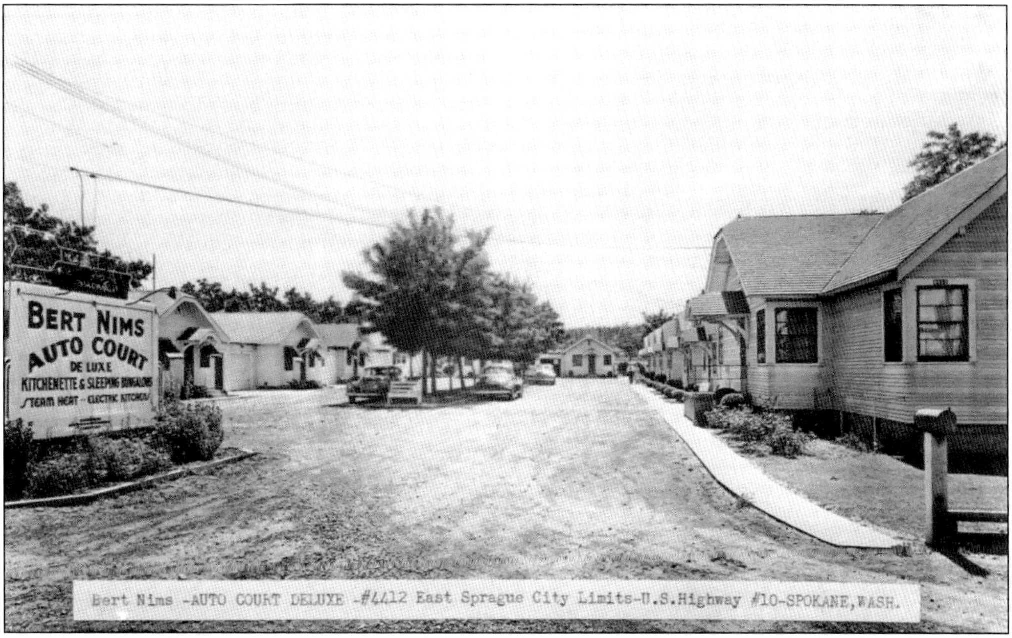

Brothers V.R. and Robert B. "Bert" Nims opened Nims Café in downtown Spokane in 1922. By 1943, Bert had branched out into the motel business with his self-named Bert Nims Auto Court at 4412 E Sprague. The motel opened with just a few units and was not finished until 1946, with two rows of bungalow-style units lining a tree-studded driveway. By 1950, it had become the Park Lane Motor Court. (WSDA, AR-28001001-ph000090.)

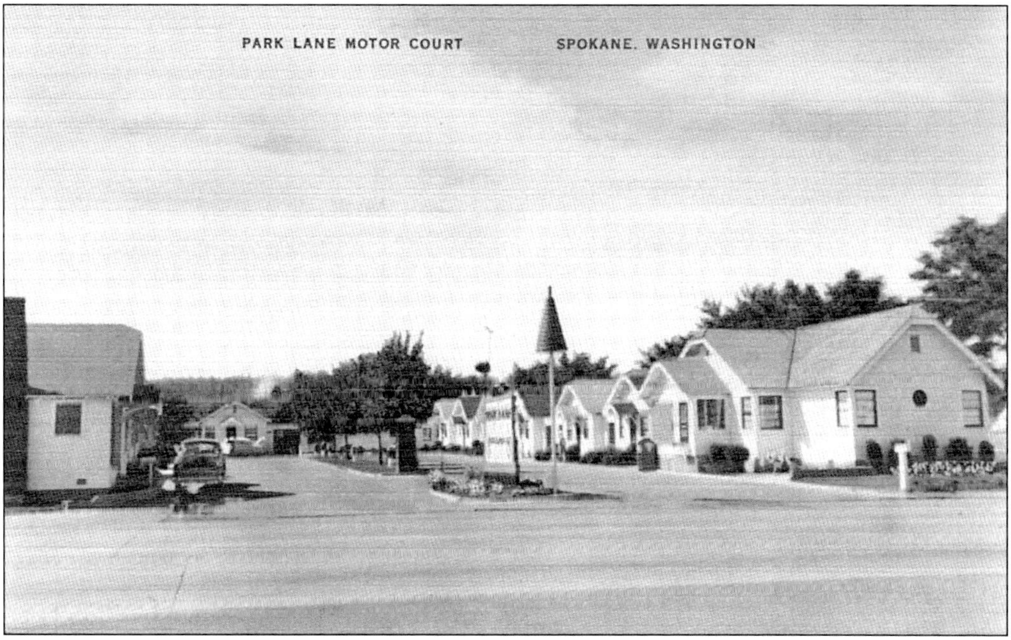

Little changed in appearance, the Park Lane Motor Court is still in business. At the time this photograph was made, the motel offered 18 sleeping bungalows and kitchenettes and was owned and operated by Mr. & Mrs. James A. Upchurch. In April 1961, Upchurch was robbed of $271 at gunpoint; the robber was apprehended in El Paso, Texas, two weeks later.

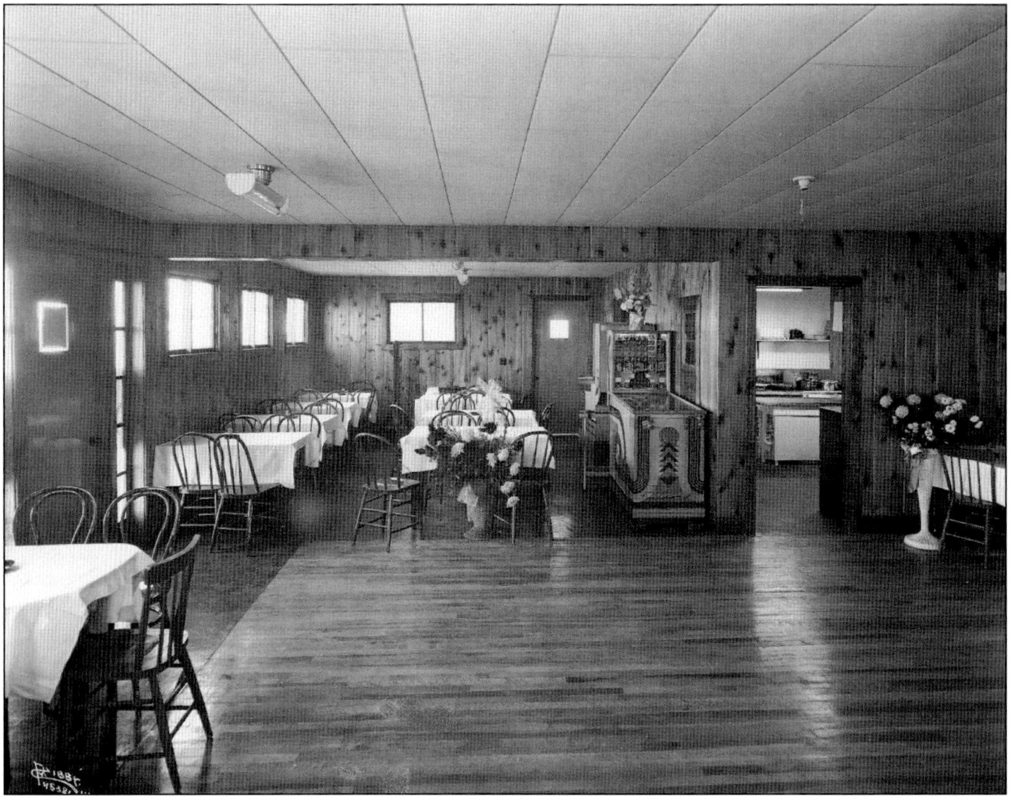

Chester's Mammy Shack was another chicken-dinner roadhouse along the highway east of Spokane. Opened by Chester Jones at E4310 Sprague in 1943, the restaurant's wood-paneled interior featured a large dance floor and a floral-decorated pinball machine. In June 1946, burglars made off with five pounds of butter, 12 chickens, and $85 in nickels, scattering 13 pounds of sugar and coffee around the place on their way out. (MAC, L87-1.45386-45.)

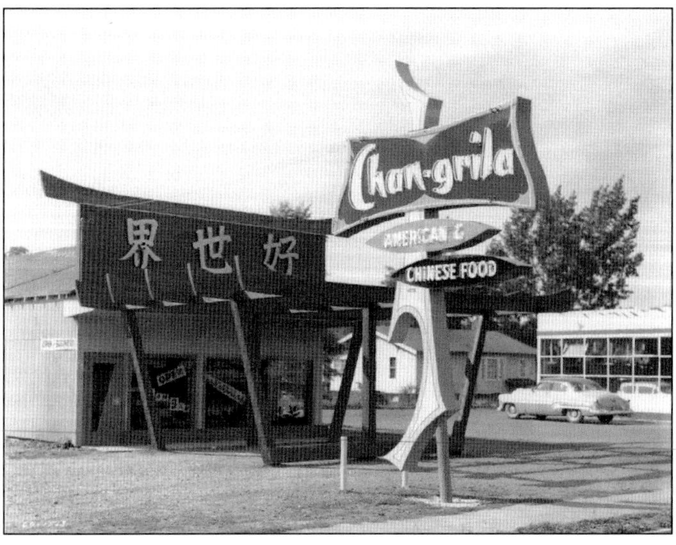

Located at E3420 Sprague, the Chan-grila (apparently a word-play on Shangri-La) was a simple frame building with an ornate façade applied to make it appear Oriental. When the restaurant became the Peking Gardens in 1974, the original sign was repurposed to advertise the new name. Destroyed by fire in February 2004, long-time owners George and Ruth Lo rebuilt and continued to operate the restaurant for several more years. (MAC, L87-1.1513-60.)

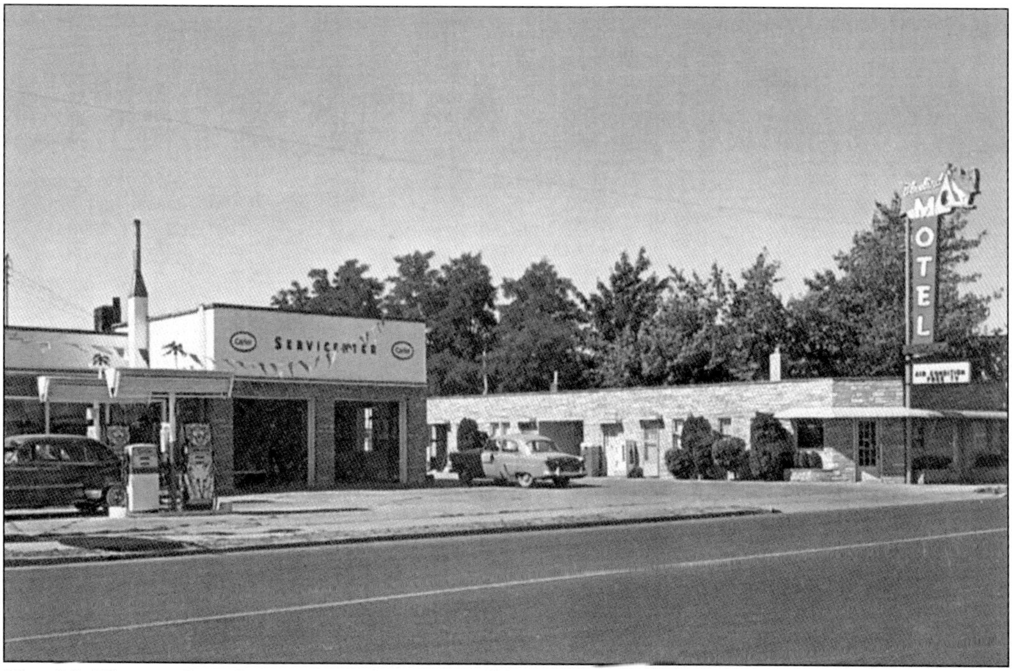

Harold's Tourist City Court opened prior to 1937 at E3320 Sprague Avenue. The Bluebird Motel was later built at the same location. The Bluebird consisted of 26 air-conditioned units, each with room phones and television, some with kitchens; a service station; a garage; and living quarters and office. Lynn and Dorothy Brownson owned the motel in 1960. Today, the site is a used car dealership. (JCO'D).

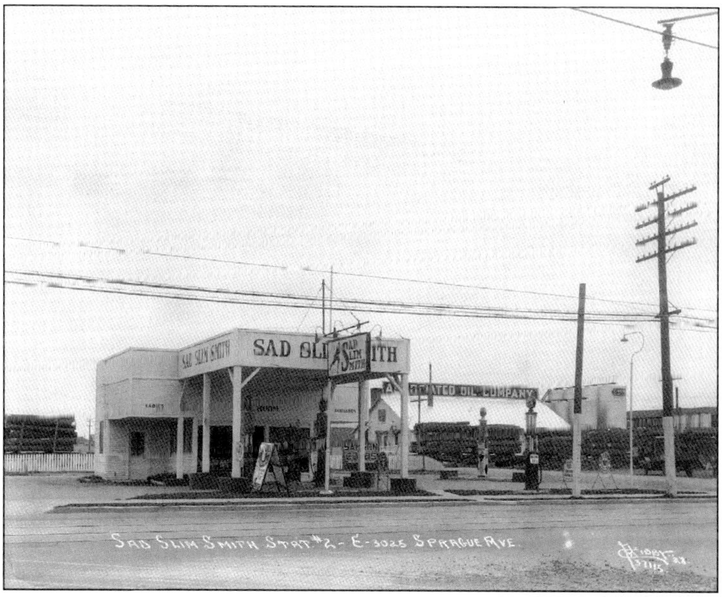

W.W. Smith owned four "Sad Slim" Smith's Tire Stations in Spokane in 1926, selling several competing brands of gasoline including Red Crown, Union, and Associated. In 1928, Smith sold Station No. 2, located at E3205 Sprague Avenue, to the Tide Water Associated Oil Company for approximately $10,000. The station went through a number of operators in the next two decades and had disappeared by 1957. (MAC, L87-1.37115-28.)

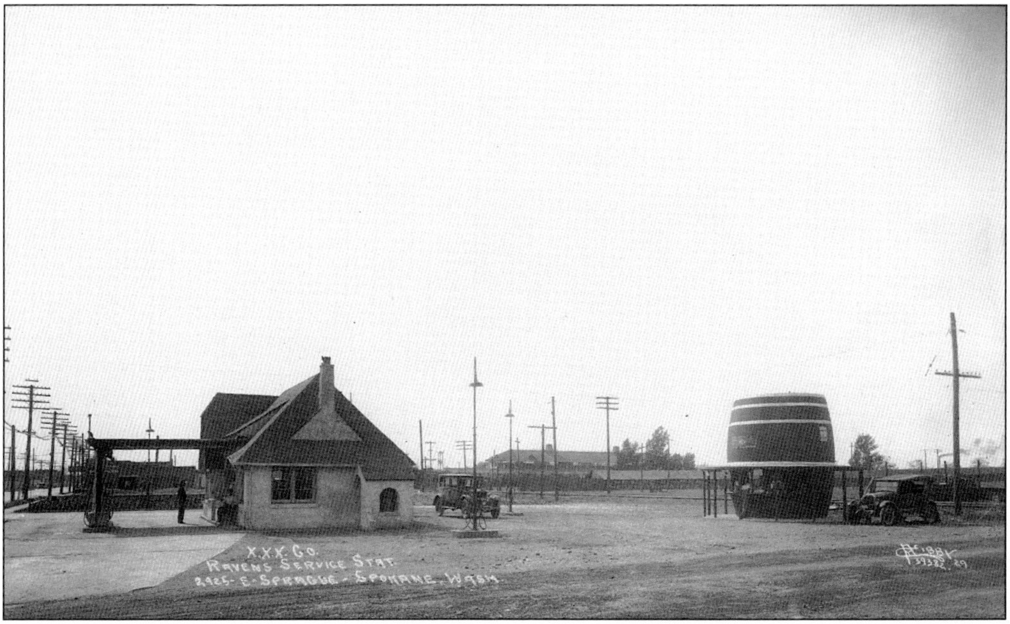

Two men take their leisure awaiting customers in the front bay of Clarence Raven's service station at E2925 Sprague Avenue. The substantial stone building, complete with chimney and a sheltered side entrance (probably to the restrooms), looked more like a residence than a place to buy gas. At the rear is a Triple XXX root beer barrel operated by Cora Graham and, in later years, Gladys Knee. (MAC, L87-1.39382-29.)

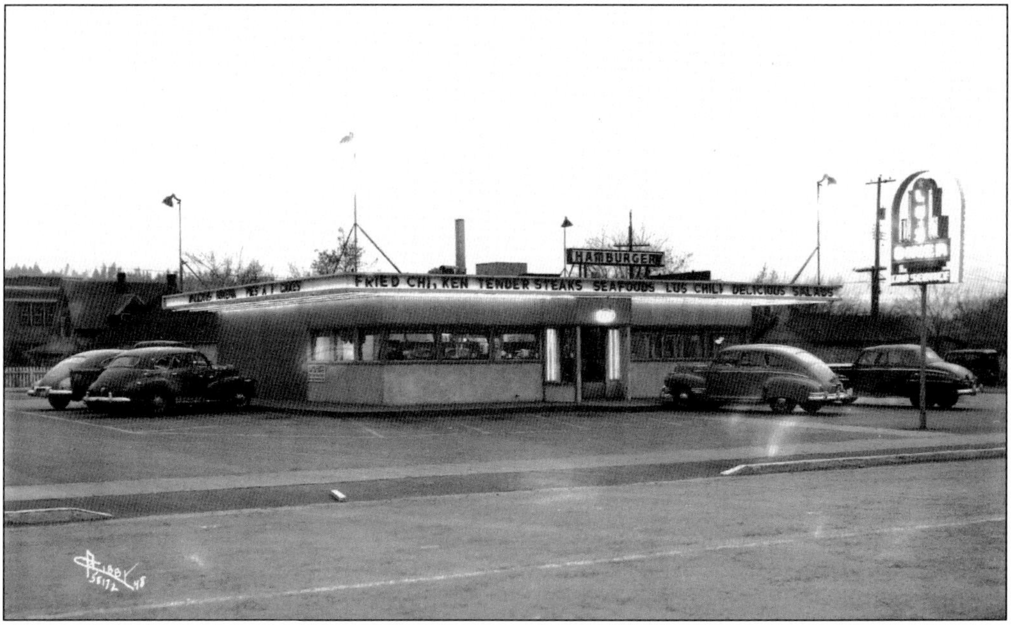

Luman "Lu" Shannon operated several Spokane restaurants and taverns during his career, opening Lu's Diner at E2904 Sprague Avenue in 1945. Lu's featured typical diner fare such as steaks, chicken, and homemade pies, and was famous for its Coney Island sandwich—a chili dog piled high with diced onions and yellow mustard. The site of Lu's is now occupied by a Vietnamese restaurant. (MAC, L87-1.58172-48.)

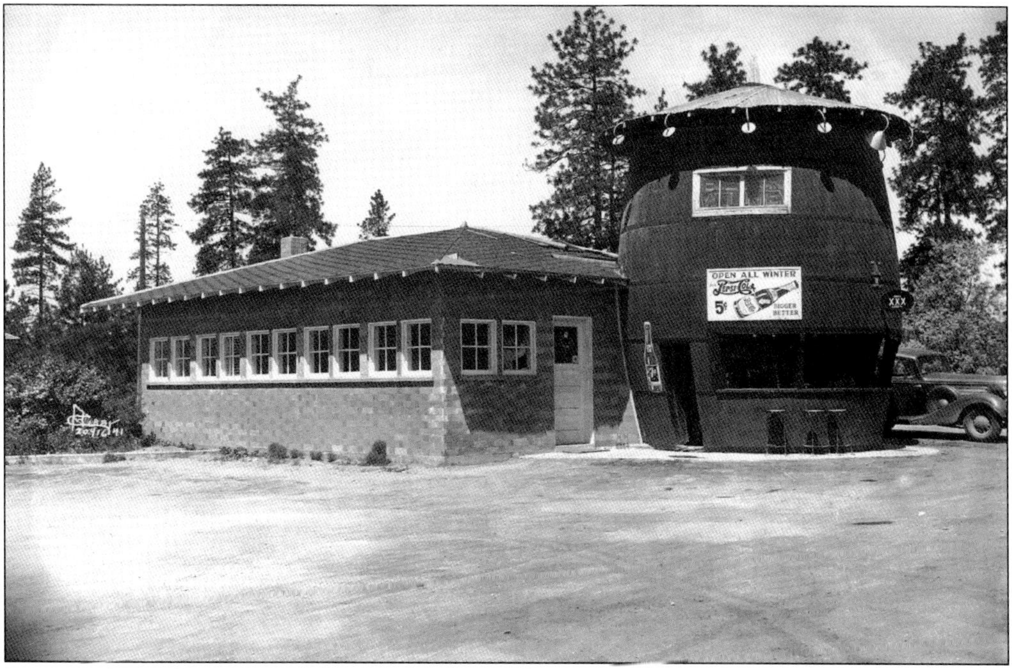

In 1928, the Galveston Brewing Company launched an aggressive marketing campaign to sell franchises for its Triple XXX root beer brand. It found a ready market in Spokane; at least eight barrel-shaped Triple XXX Thirst Stations opened over the next few years in various parts of the city. This example was located at E 2125 Sprague and, in later years, sold beer and wine in addition to root beer. (MAC, L87-1.20916-41.)

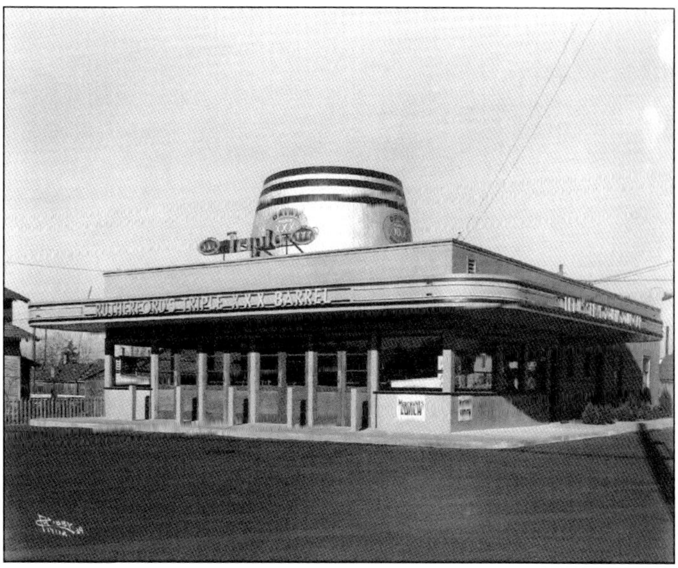

The Rutherford family opened their first Triple XXX outlet in Renton, Washington, in 1930. By 1941, the Rutherfords had developed a distinctive design for their restaurants—a streamlined, neon-striped building surrounding a giant silver barrel displaying the Triple XXX emblem—and opened three additional outlets in Seattle and two in Spokane including this one, an updated replacement of the barrel shown in the previous photograph. (MAC, L87-1.17112-39.)

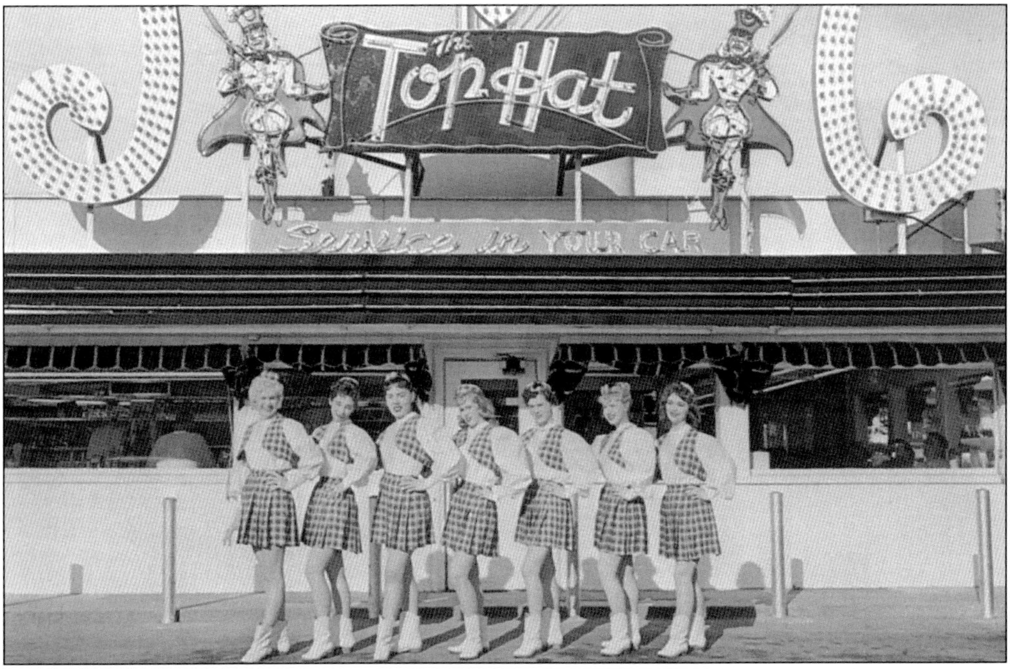

Short-skirted carhops display their style in front of the Top Hat Fountain Café, an all-night restaurant at E 2101 Sprague Avenue. When this photograph was taken, the Top Hat had been open for 38 years. Service was available in one's car, with trays attached to rolled-down windows or inside the restaurant. Newspaper reports indicate the Top Hat was a favorite spot for "cruisers" and occasionally saw altercations between rival groups. (JCO'D.)

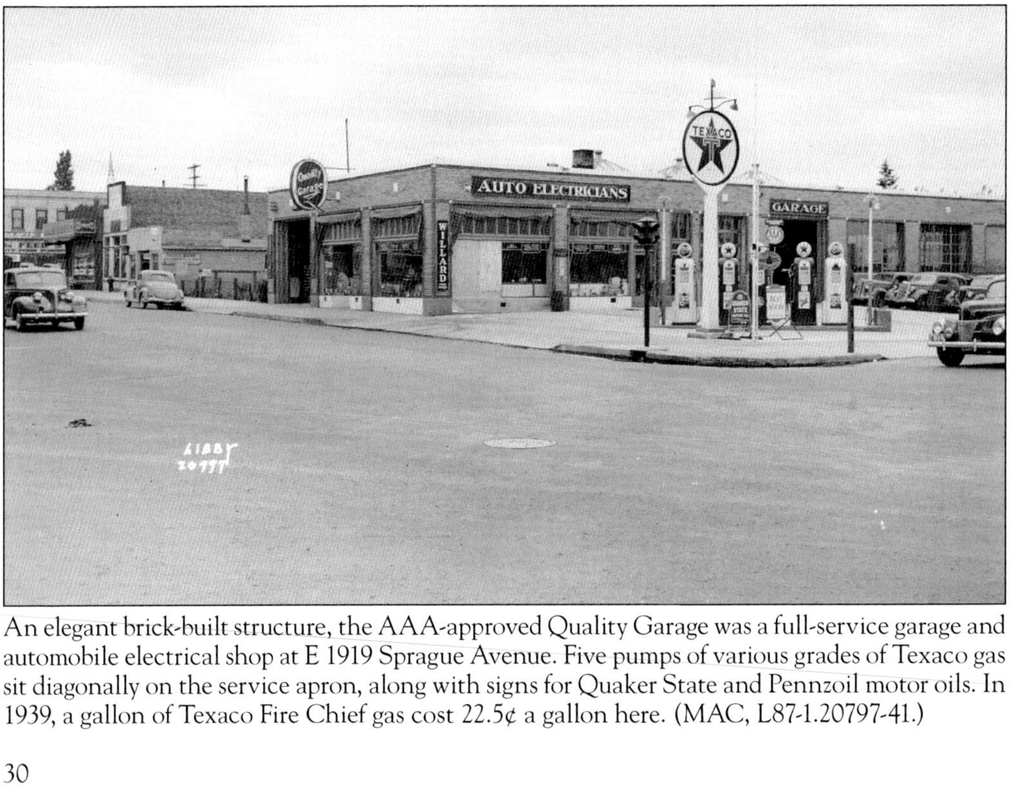

An elegant brick-built structure, the AAA-approved Quality Garage was a full-service garage and automobile electrical shop at E 1919 Sprague Avenue. Five pumps of various grades of Texaco gas sit diagonally on the service apron, along with signs for Quaker State and Pennzoil motor oils. In 1939, a gallon of Texaco Fire Chief gas cost 22.5¢ a gallon here. (MAC, L87-1.20797-41.)

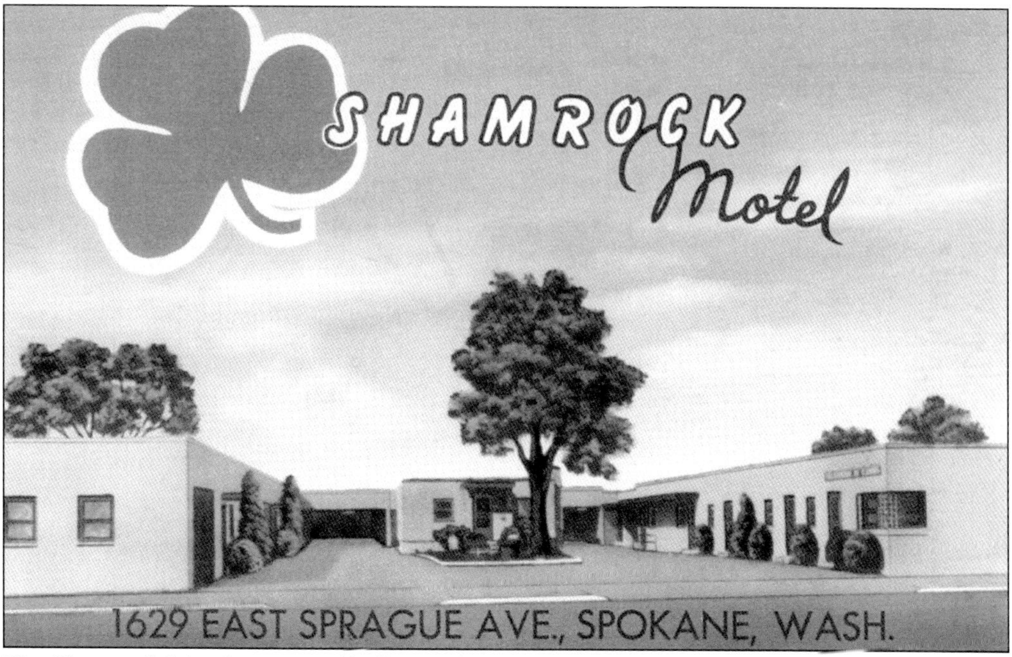

The Shamrock Motel was a small affair with nine stucco-coated units, all with kitchenettes and hot water heat, and a small green space just big enough for a tree and a few lawn chairs. Originally owned by Mr. and Mrs. Allen Daniel and later by Raymond Lynch, who was also associated with the Park Lane Motel, it was located at E 1629 Sprague Avenue. The site is now a parking lot.

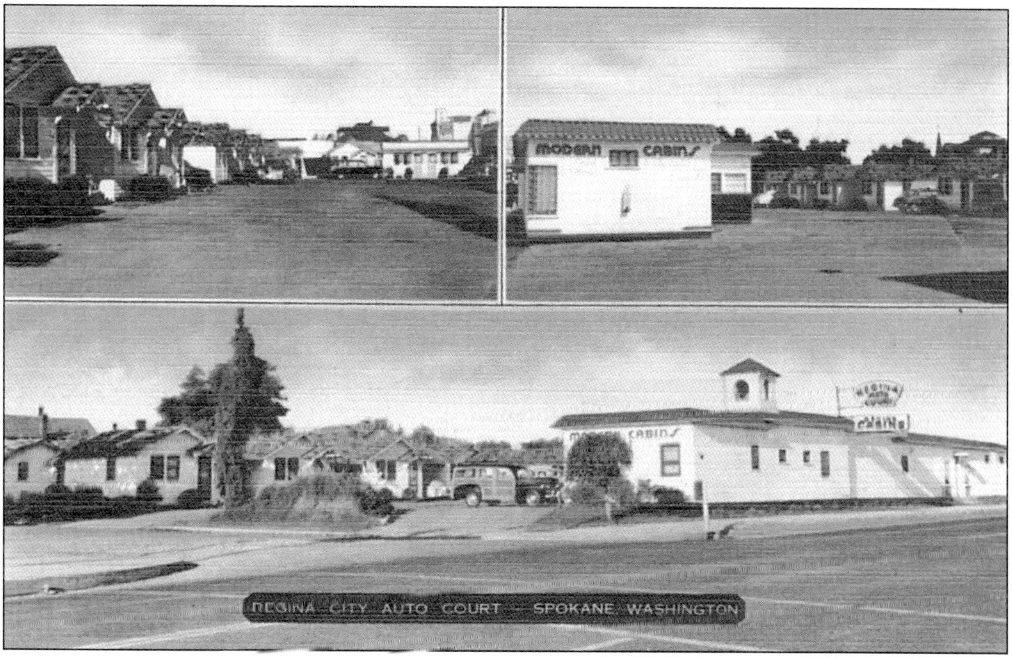

Harry Frazier opened the Regina City Tourist Camp at E 1618 Sprague in 1931. Owned by Paul Roseberry in 1947 and renamed the Regina City Auto Court, the 48 units were a combination of styles: older detached wood cottages and a newer stucco building with a continuous block of rooms. The address hosts a post office building today.

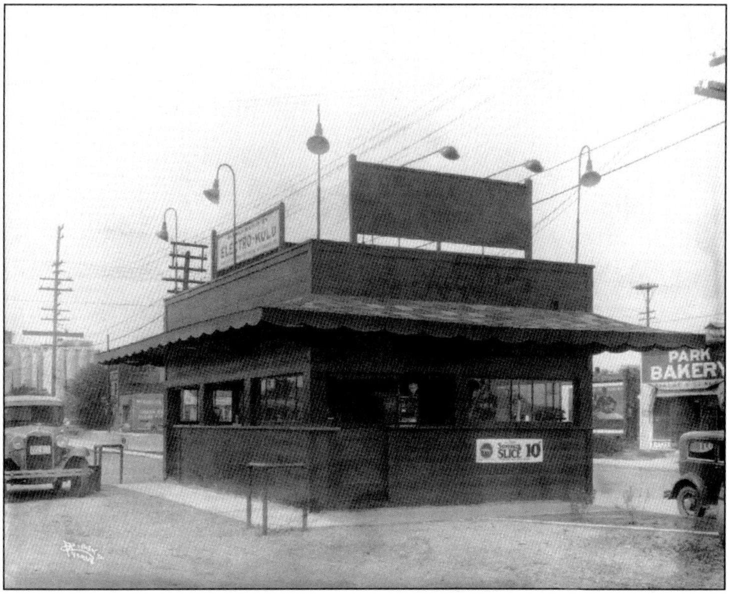

Triple XXX was not the only root beer in Spokane. A local brand, Regina, "Queen of Drinks," had a stand at the Regina City Tourist Court that sold its root beer along with popcorn, peanuts, and Chocolate-Coated Sunfreze Slice. Regina had other Spokane outlets at N 1433 Division Street, N 4619 Market Street, and the Riverside Food Shop, as well as stands in Walla Walla and Yakima. (MAC, L87-1.42468-30.)

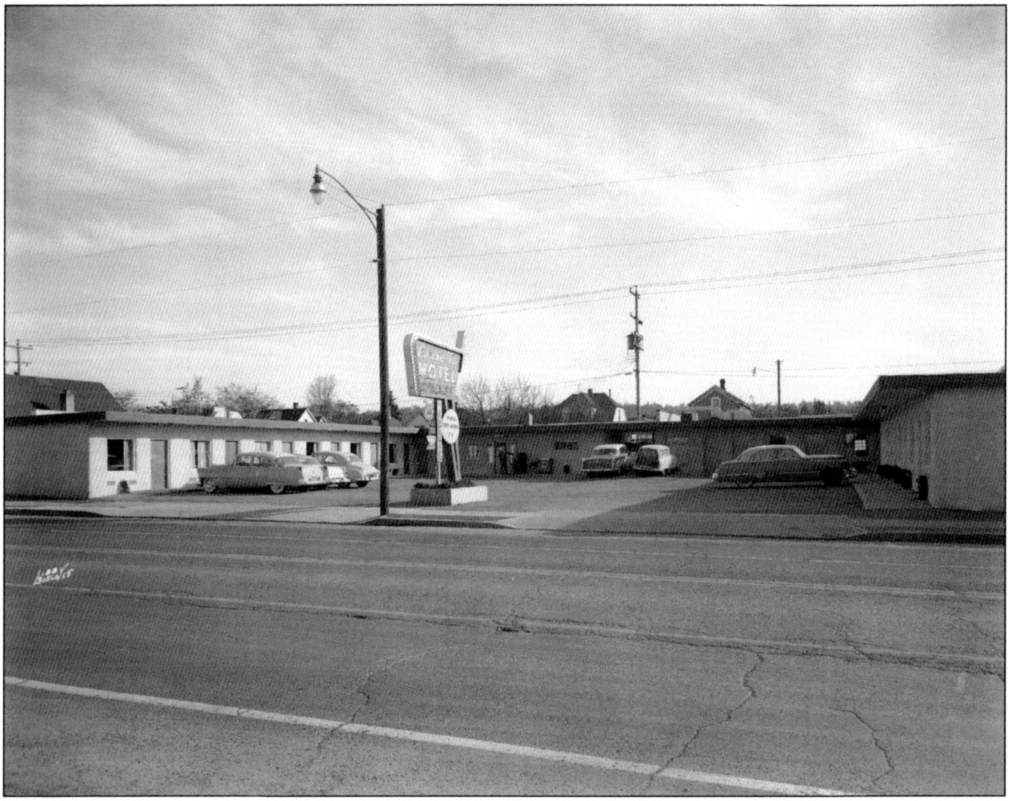

In 1952, Virgil Carroll was operating the Spokane Lodge; by 1957, he had opened his own motel, Carroll's, at E 1234 Sprague Avenue. With 14 units, Carroll's was a rather severe-looking square-U configuration built around a paved parking area. In later years, it was known as the Budget Saver. Today, the building is occupied by the Union Gospel Mission Crisis Center for Women and Children. (MAC, L87-1.90560-58.)

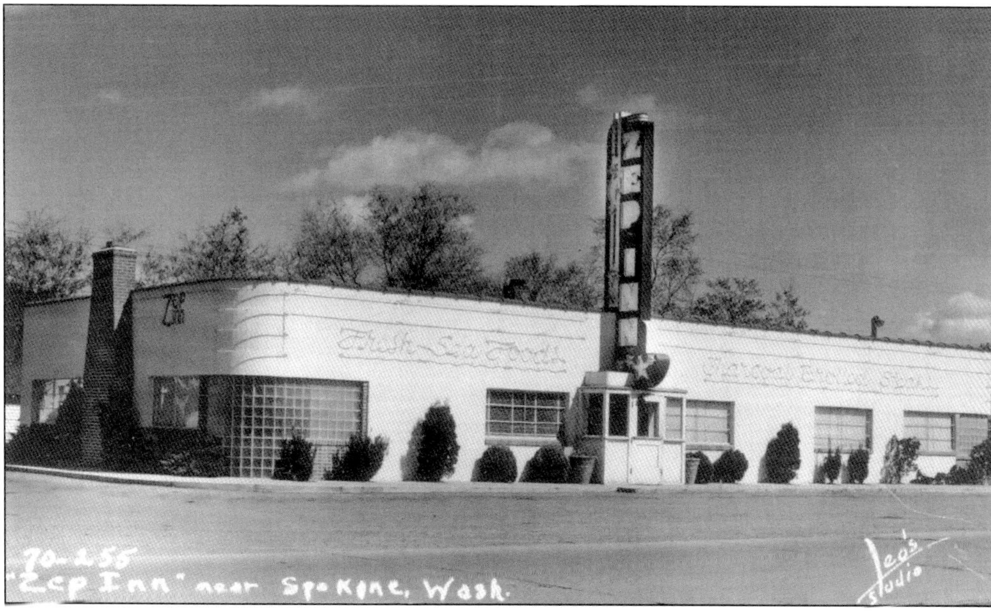

Pierce and Bernice Harbottle owned and operated the Zep Inn from 1932 to 1972. The inn, housed in a streamlined Art Deco building, was located at E 5207 Trent, a route occasionally designated as Highway 10-A. Steak and chicken dinners were featured, with cocktails in the Wishing Well lounge; the Mexican Room was available for dinners and parties. Today a Chinese restaurant occupies the site.

Jack and Frank Knight opened Knight's Diner in 1931 at E 5615 Trent Avenue across from the Parkwater rail yard in a recycled railroad passenger car. Frank moved to Seattle in 1938 to open another Knight's Diner there. The Spokane Knight's moved to Division Street before coming to rest at 2909 N Market Street. The Seattle Knight's eventually ended up in Spokane, where it is known as Frank's Diner. (MAC, L87-1.10774-37.)

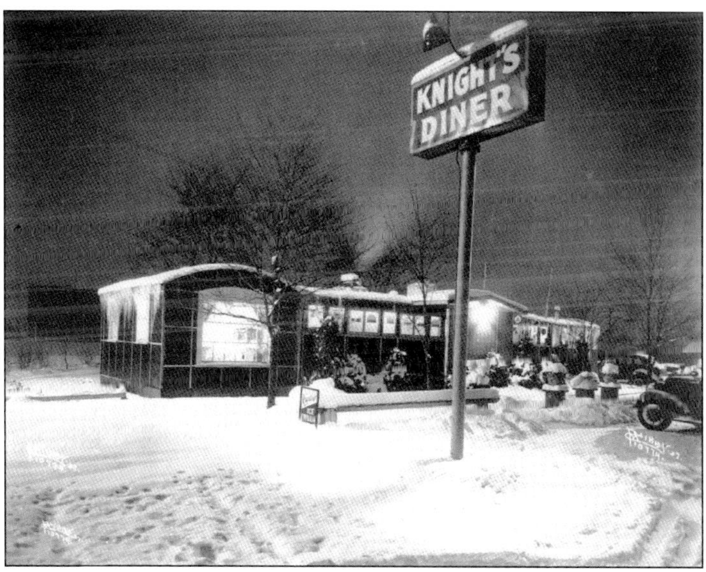

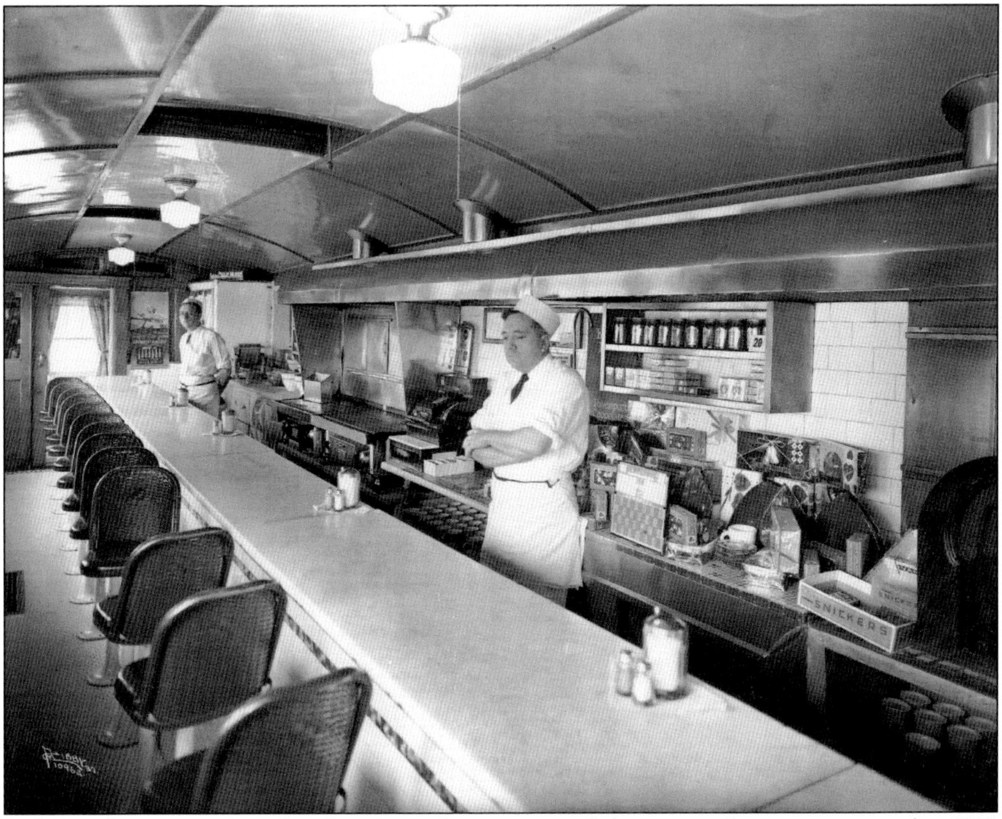

Jack Knight contemplates his diner's spotless counter while waiting for customers in this 1937 view; his unidentified cook stands ready at the flat-top griddle. The rattan-backed counter stools are an interesting feature. On the counter behind Knight are a variety of candy boxes; labels for Snickers, Davenport Chocolates, Allen's Nut Toffee, and Marie Antoinette Valentine and Craft Selection candies can be seen. (MAC, L87-1.10962-37.)

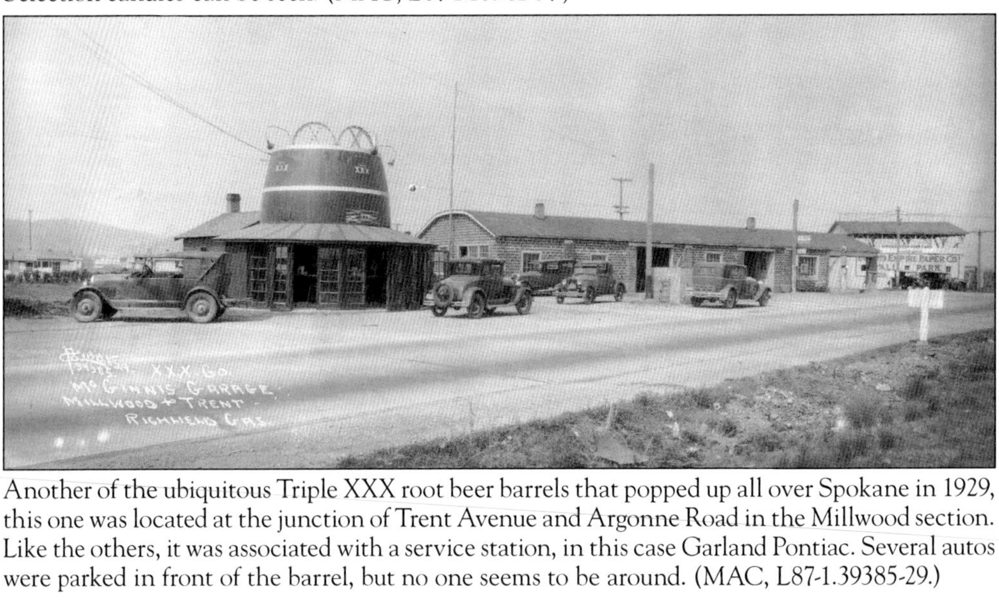

Another of the ubiquitous Triple XXX root beer barrels that popped up all over Spokane in 1929, this one was located at the junction of Trent Avenue and Argonne Road in the Millwood section. Like the others, it was associated with a service station, in this case Garland Pontiac. Several autos were parked in front of the barrel, but no one seems to be around. (MAC, L87-1.39385-29.)

Two

Spokane to Cle Elum via Waterville and Wenatchee

The Sunset Highway followed Sprague Avenue through downtown Spokane, then shifted a few blocks south onto Sunset Boulevard. Just west of the city, the highway crossed over Latah Creek, passed a few remaining roadside businesses, and climbed 400 feet to the rim of the Columbia Plateau, which the highway traversed for the next 120 miles. The route passed through the following towns spaced roughly 30 miles apart like beads on a necklace: Reardan, Davenport, Creston, Wilbur, Almira, Hartline, Coulee City, and Waterville. This was (and still is) wheat country. Farmhouses are few and far between; grain elevators look like skyscrapers in the distance.

Beyond Waterville, the highway dropped 2,000 feet through Pine Canyon to the Columbia River at Orondo, then swung south 30 miles to cross the river into Wenatchee. The largest city between Spokane and Seattle, Wenatchee offered a full complement of services with hotels, restaurants, and auto repair shops.

From Wenatchee, the Sunset Highway followed the same-named river west for 15 miles to Dryden; a few miles beyond, the highway turned south. The Sunset Highway, as did the Yellowstone Trail before it, followed along Peshastin and Ingalls Creeks to scenic but problematic Blewett Pass. Often closed by winter snows, Blewett Pass plagued motorists for years. Eventually, the Sunset Highway was rerouted to the south and completely bypassed Blewett Pass. In later years, the Blewett Pass route itself was relocated and a modern highway—Highway 97—constructed five miles to the east. Confusingly, the new route over Swauk Pass is referred to as the Blewett Pass Highway.

South of Blewett Pass the Sunset Highway made a junction with the Inland Empire Highway at Virden, from which point motorists could continue west to Cle Elum or south to Ellensburg and Yakima (the route of today's Highway 97). Beyond the junction, the Sunset Highway contoured the edge of the Teanaway River valley into Cle Elum.

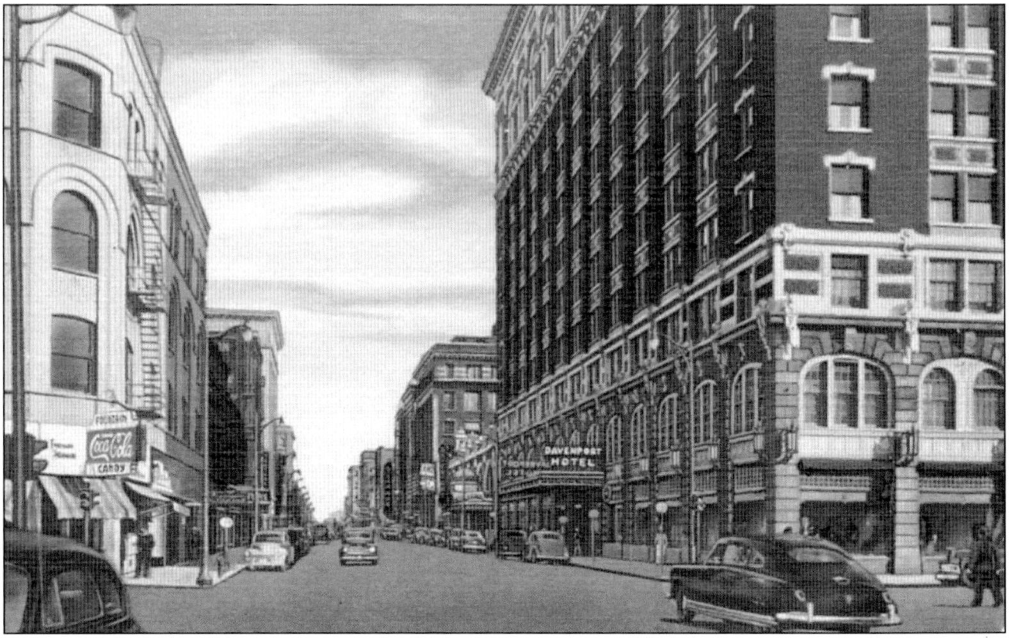

With a 1940 population of 122,000, Spokane was the largest city westbound motorists had encountered since Minneapolis. Its big-city bustle would have been attractive to those craving creature comforts after 1,000 miles of small towns and wide-open spaces. This view looks east along Sprague Avenue, the route of the Sunset Highway through the business district, and prominently features the famous Davenport Hotel.

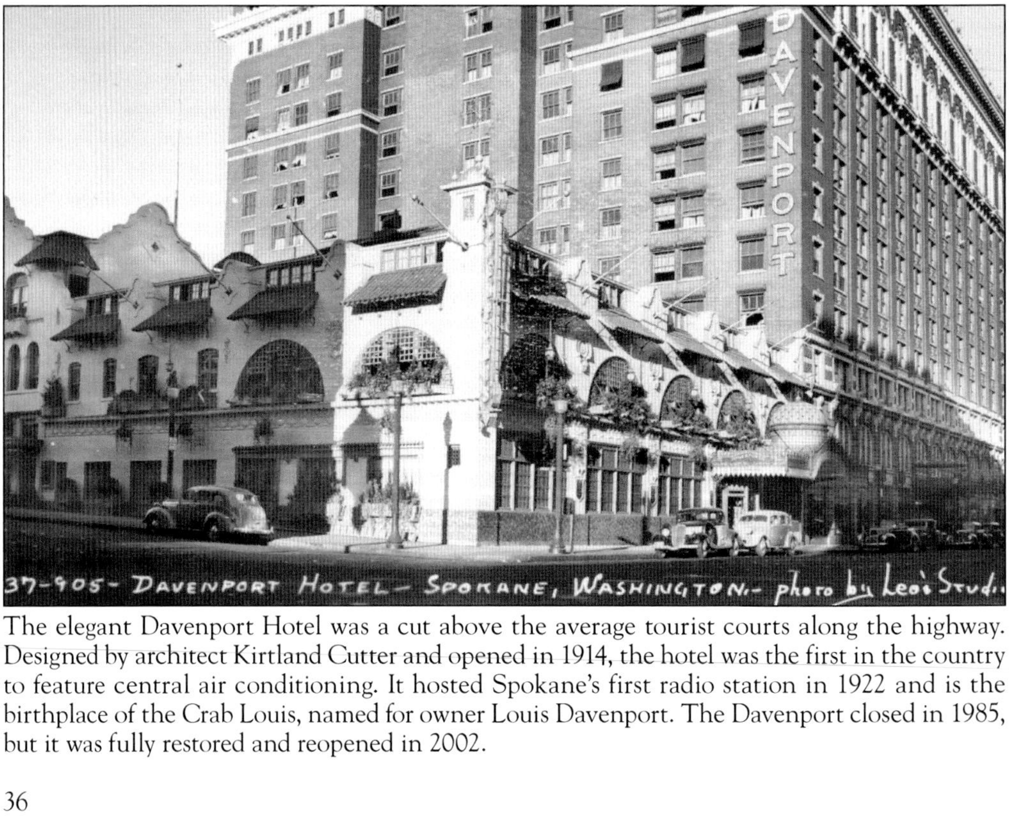

The elegant Davenport Hotel was a cut above the average tourist courts along the highway. Designed by architect Kirtland Cutter and opened in 1914, the hotel was the first in the country to feature central air conditioning. It hosted Spokane's first radio station in 1922 and is the birthplace of the Crab Louis, named for owner Louis Davenport. The Davenport closed in 1985, but it was fully restored and reopened in 2002.

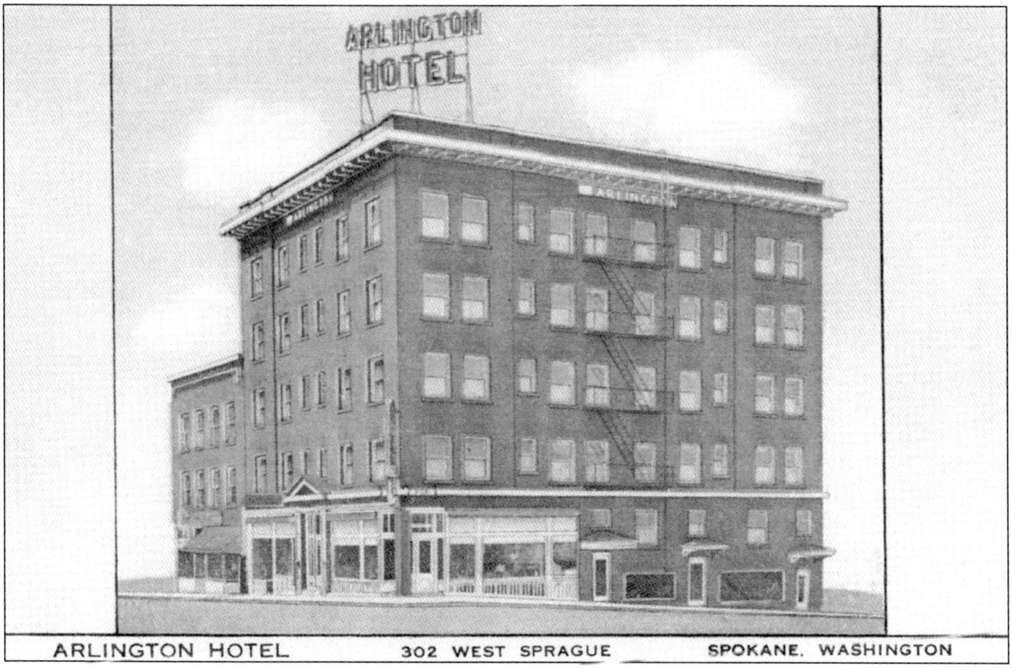

Down the street from the Davenport was the Arlington Hotel, more typical of the big-city accommodations available to travelers unable to afford the Davenport—in 1920, the Arlington's rooms were 75¢ nightly. The five-story hotel was owned by H.C. Lewis and had 121 rooms, with a lobby and commercial spaces on the ground floor. The building was heavily damaged by fire in 1999 and subsequently demolished. (JCO'D.)

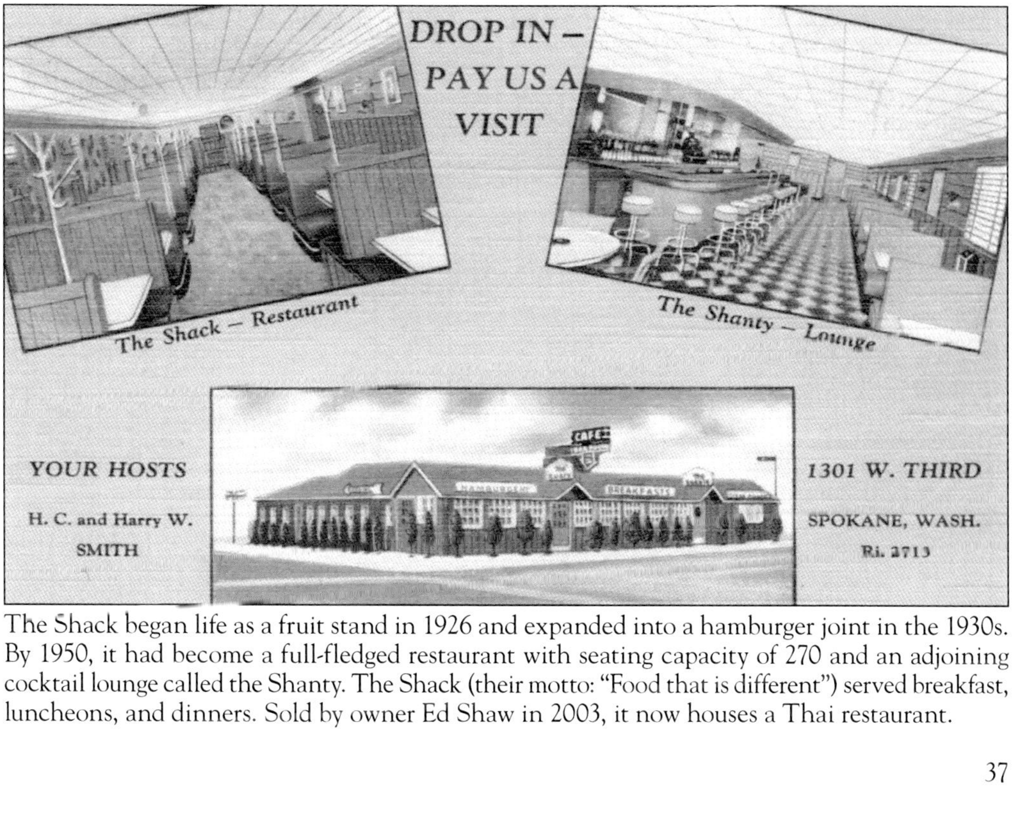

The Shack began life as a fruit stand in 1926 and expanded into a hamburger joint in the 1930s. By 1950, it had become a full-fledged restaurant with seating capacity of 270 and an adjoining cocktail lounge called the Shanty. The Shack (their motto: "Food that is different") served breakfast, luncheons, and dinners. Sold by owner Ed Shaw in 2003, it now houses a Thai restaurant.

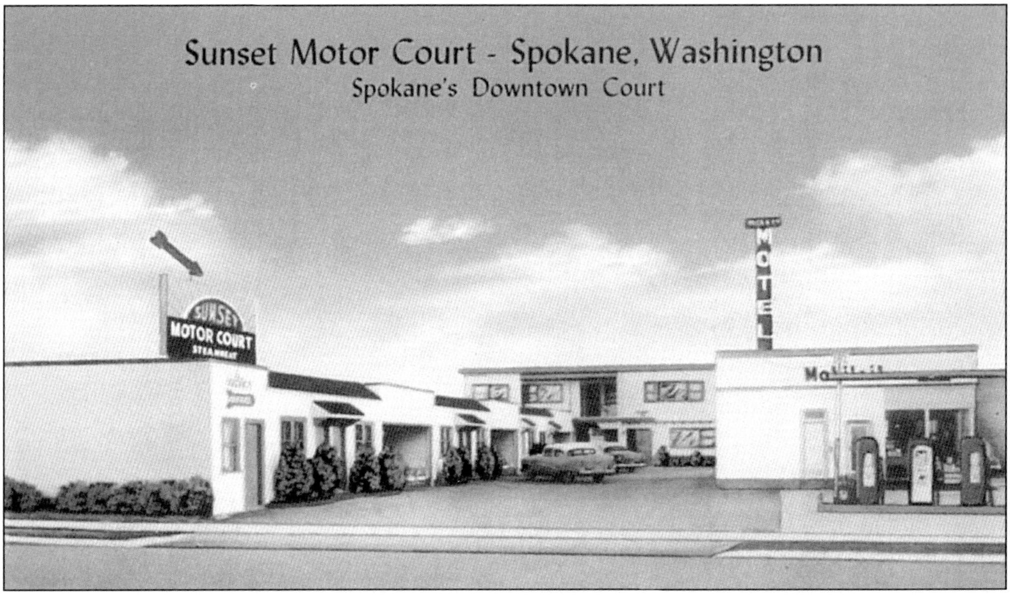

Sunset Motor Court was a block off the main highway at W 1518 1/2 3rd Avenue. The 11-unit motel opened sometime prior to 1943 under Harold C. Lines and was purchased in 1948 by Stan Leland, who also owned the adjacent Mobil gas station. Nightly rates that year were $1.75–$2.75 for two people and $2.50–$4.00 for four. By 1957, the motel had added six more rooms, and Joseph G. Johnsen had become manager.

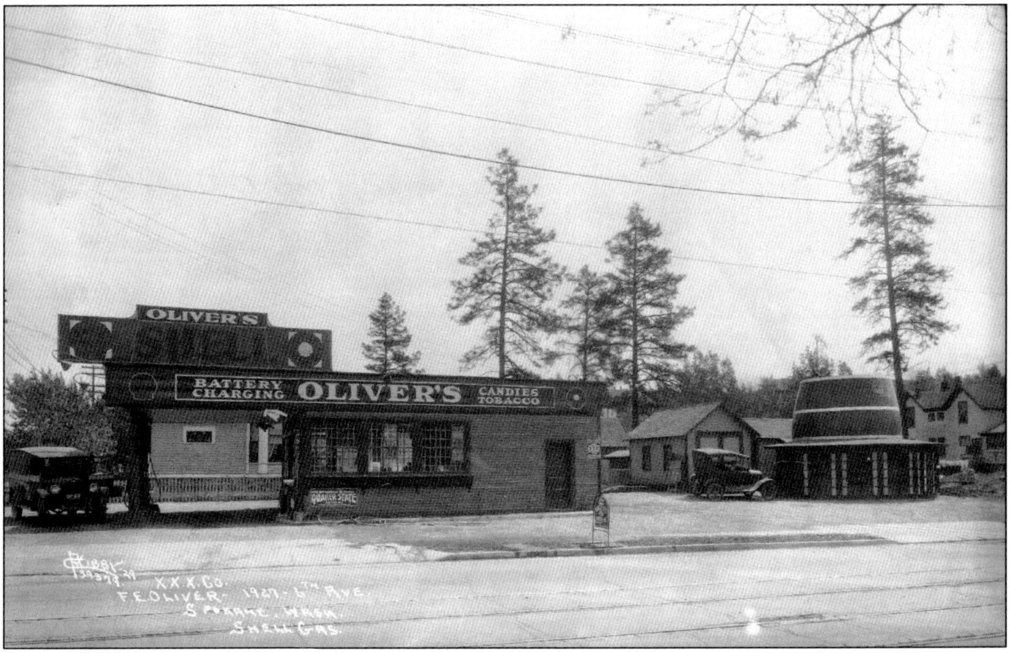

Shell Oil salesman Floyd Oliver went into business for himself by purchasing this gas station from Oscar Stowell in 1928. A Triple XXX barrel was added the following year. Oliver kept a small stock of sundries in his station, an early version of the modern mini-mart. The 1934 Spokane business directory indicates that Oliver later opened a tourist camp on the site, at W 1927 6th Avenue. (MAC, L87-1.39379-29.)

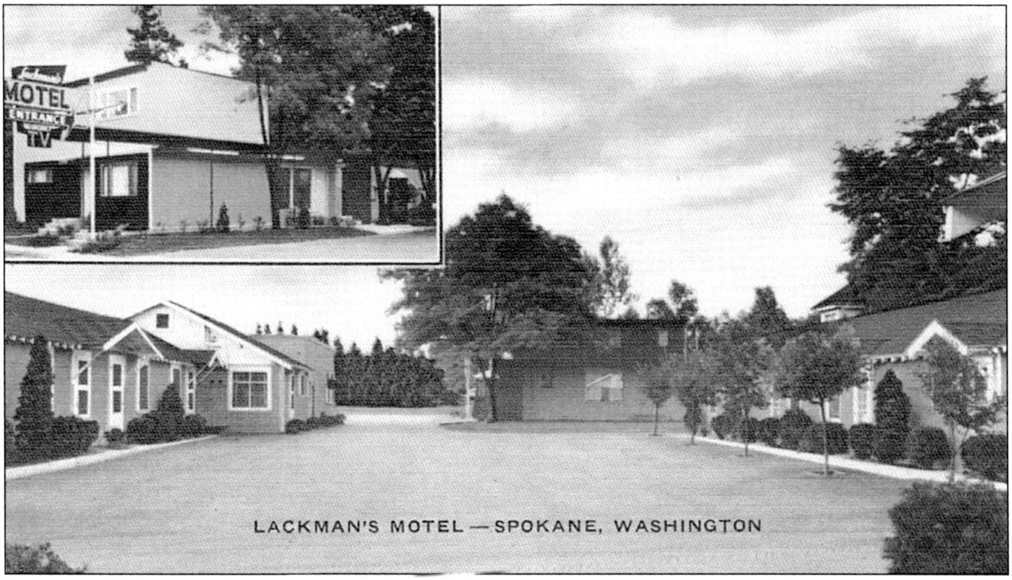

Lackman's was another off-highway motel on Spokane's west side. Opened by realtor Henry Lackman in 1947, it continued to operate into the 1950s. Lackman's consisted of 22 "all modern deluxe units," each equipped with kitchenettes, tile baths, and televisions. The motel was located in a tree-lined residential area—until Interstate 90 was built through it. The site, at W 1926 6th Avenue, is now buried under the freeway.

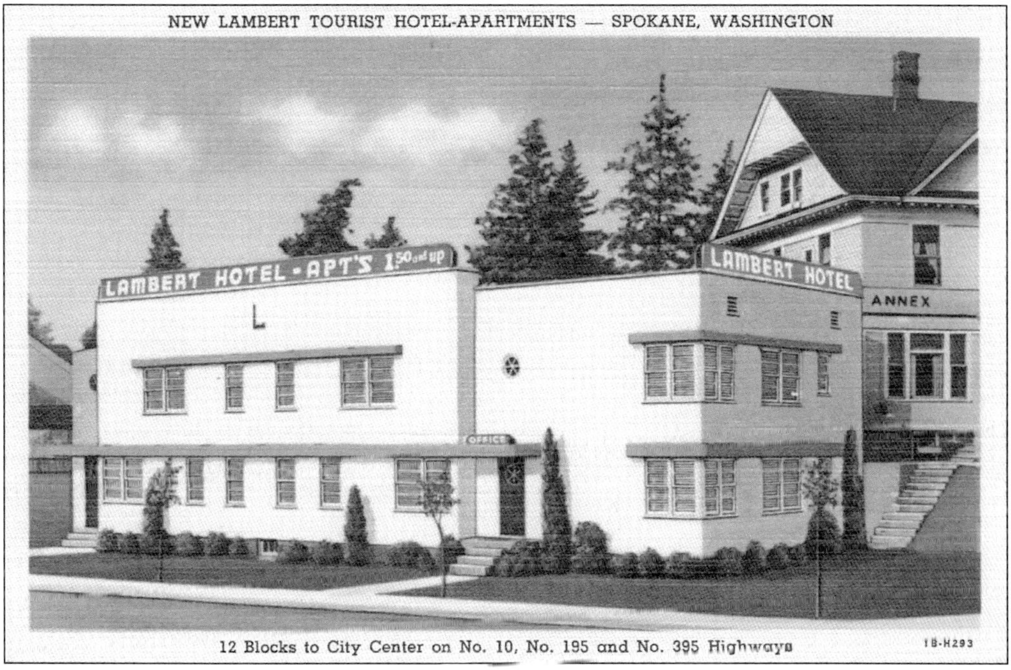

In 1942, I. J. and L.F. Lambert, proprietors of the Lambert Auto Court in Dishman, completed construction of this new hotel on the Sunset Highway west of downtown Spokane. Advertised as "strictly modern and beautifully furnished" with baths and showers in every room, the hotel was sold to Mr. and Mrs. Crawford in 1948 for $90,000. The building still stands, though not in use as a hotel. (JCO'D.)

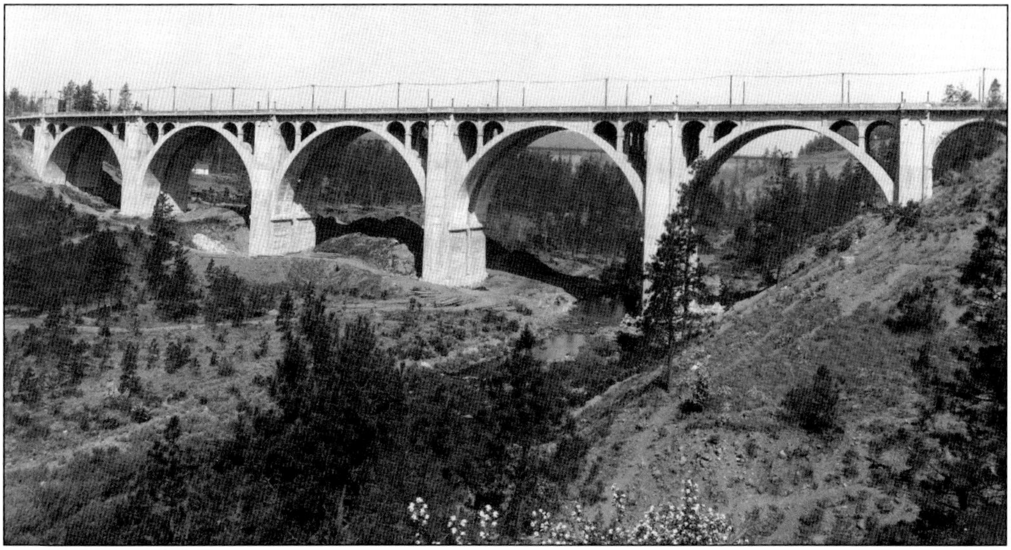

The Latah Creek Bridge pre-dates the Sunset Highway. Constructed in 1913 to connect downtown Spokane to the western uplands, the 1,070-foot-long bridge is an elegant structure consisting of seven Roman-style arches supporting the roadway and decorated end-to-end with concrete side rails and balusters. Originally four lanes wide, usage has been restricted to a lane in each direction due to the bridge's age. Plans for rehabilitation are in progress. (WSDA, 995-0329.)

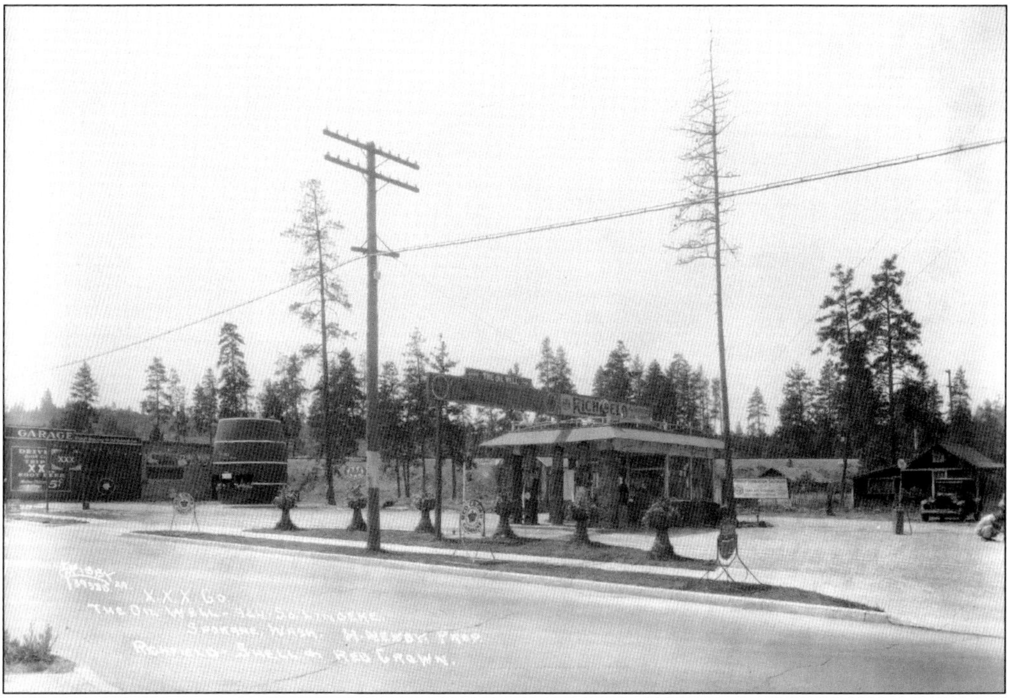

The Oil Well was in business as early as 1929. As was common at the time, the station sold several different brands of gas—Red Crown, Shell, and Richfield. Decorative plantings lined the station's driveway. Next door was a full-service garage, with an early Triple XXX root beer barrel in the lot between them. The location was S 924 Lindeke Street, two blocks off the Sunset Highway. (MAC, L87-1.39380-29.)

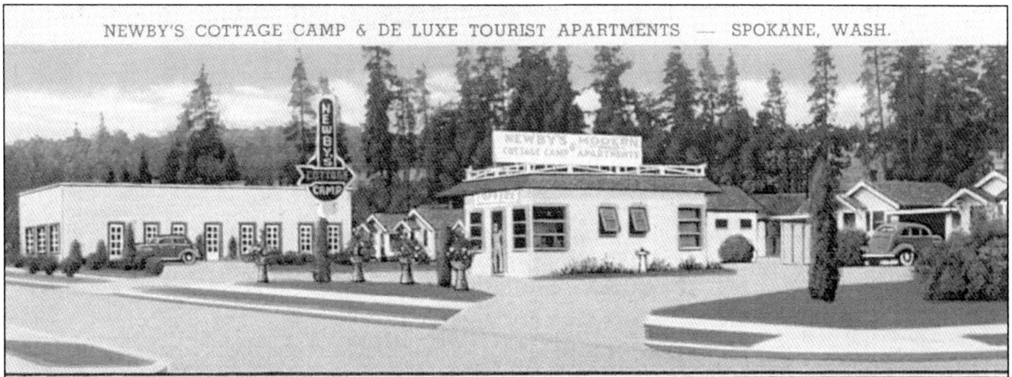

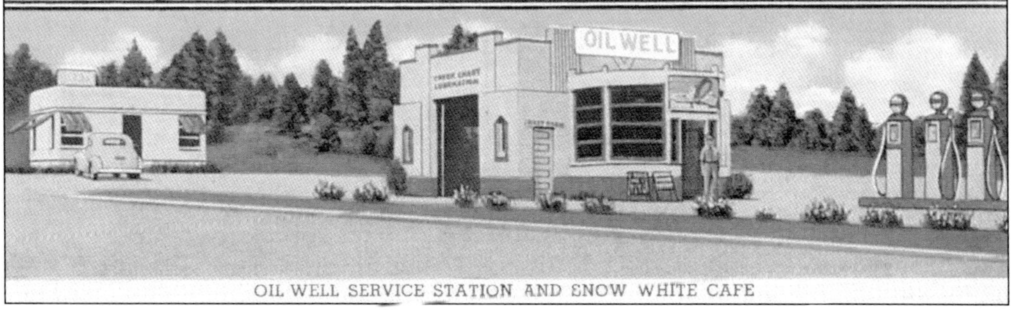

By 1942, James E. and Effie M. Newby had opened Newby's Cottage Camp on the Oil Well's site, with the old station building remodeled into an office. Individual cabins went up first; a building of deluxe apartments came later. The Snow White Cafe and a new Oil Well Service Station were adjacent. The Newbys' son Harold assumed ownership in 1955. By 1959, it had become Wynne's Motel.

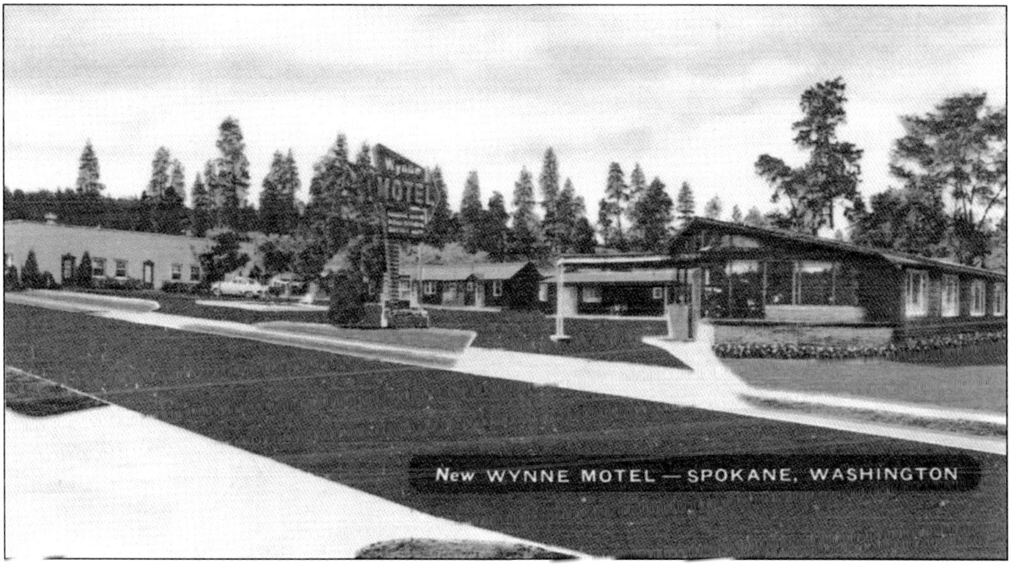

Wynne's maintained its earlier appearance for a while, but by 1962, it had been dramatically updated. The central courtyard office was gone, a large neon sign had taken its place, and a new glass-fronted office occupied a corner of the property. The cabins and apartment building, however, remained. The site of Wynne's is now a vacant lot although a successor—the West Wynn—operates a few blocks away.

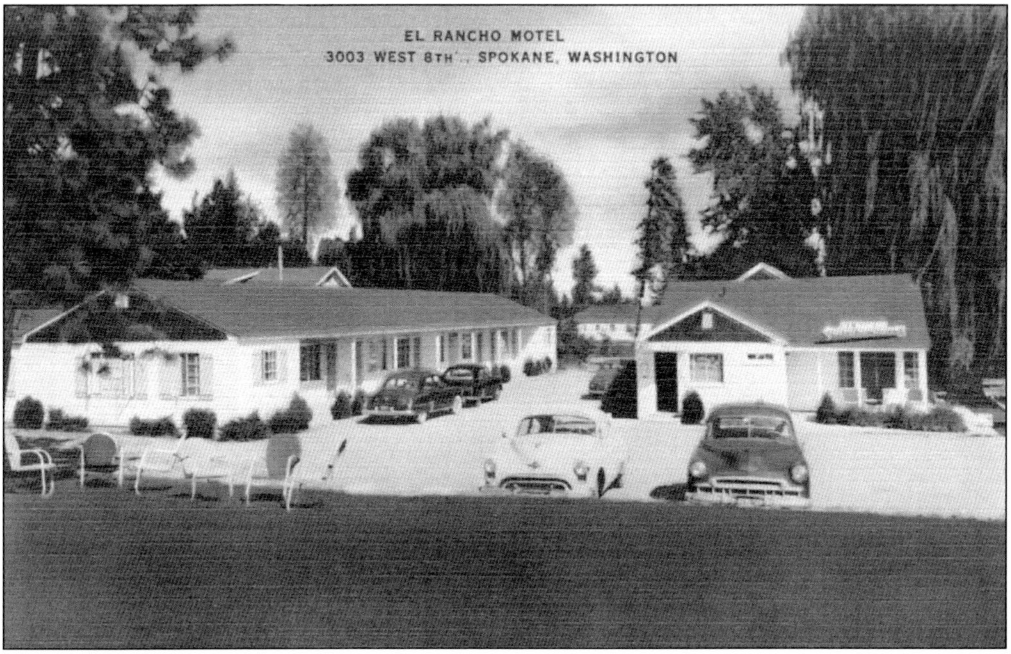

The El Rancho, at 3002 W 8th Avenue west of the Latah Creek Bridge, was a shady spot at the foot of the Sunset Highway's climb out of Spokane. The 13-unit motel was in business in 1948 under ownership of Mr. and Mrs. Fred Barnette. In 1951, Mr. and Mrs. H.E. Klundt purchased the El Rancho for $98,000 and promptly took over operation. It is no longer in business.

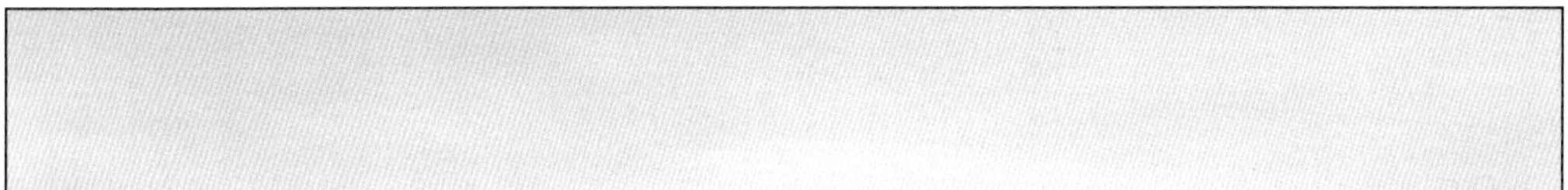

The Sunset Highway is empty of vehicles in this view looking east toward Spokane from the hills above the city. An unusual feature is the wide shoulders, which appear to be unpaved and may date the photograph to shortly after highway reconstruction in the 1930s. Later views show extra paved lanes in each direction atop the shoulders—and much more traffic.

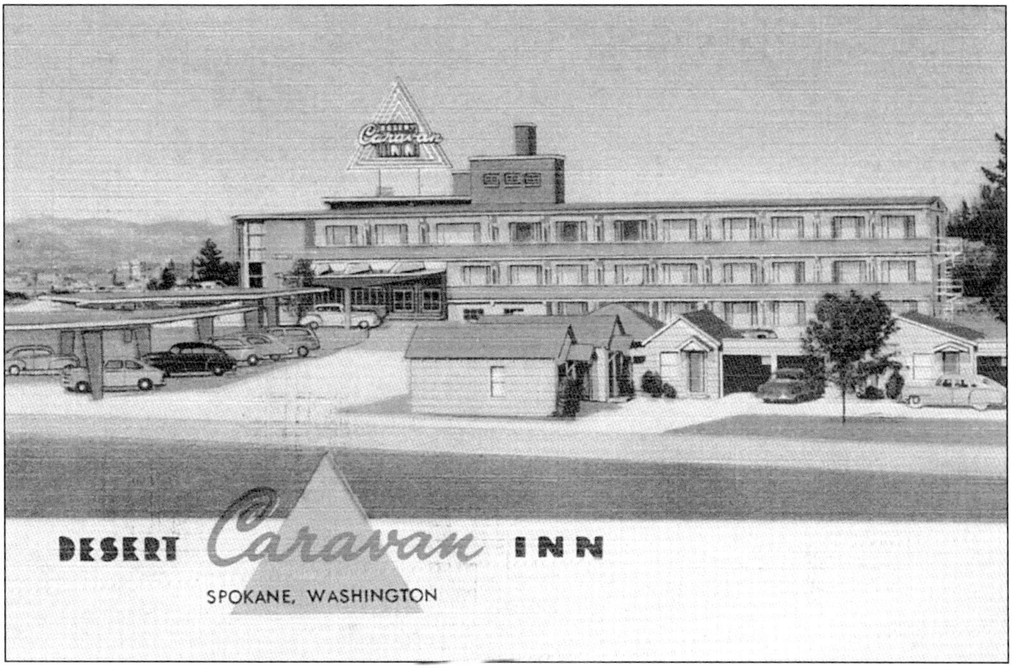

The Desert Caravan Inn was situated at the top of the climb out of Spokane. The inn boasted of its location high among the pines, with an outstanding view from its dining room out over Spokane and the mountains to the east. Two generations of motel architecture can be seen in this view—early detached cabins and 1950s-style modern.

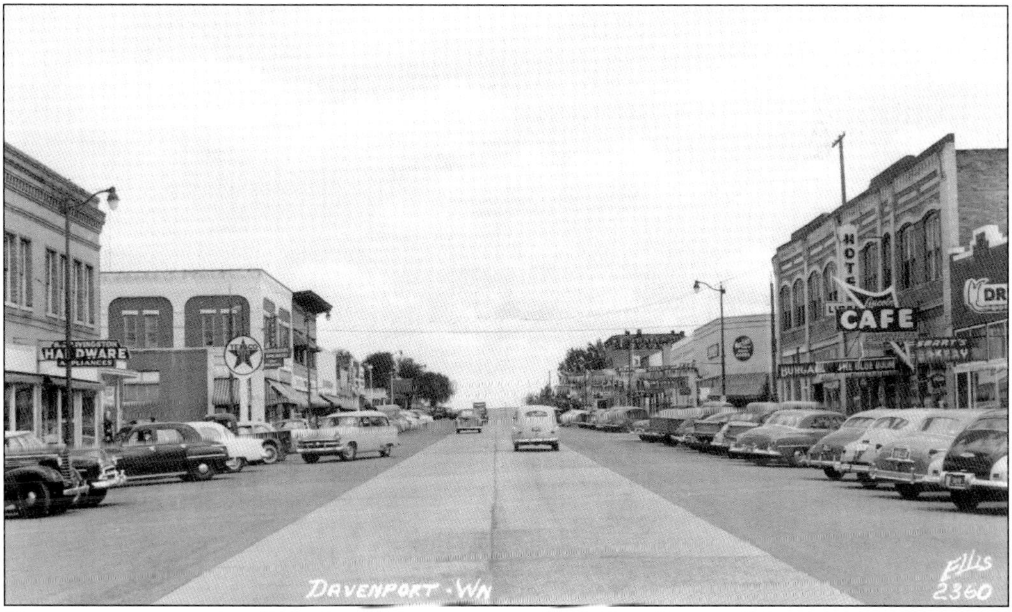

Davenport was, and still is, the largest town in the 160-mile stretch between Spokane and Wenatchee. Its status as seat of Lincoln County and a major shipper of wheat ensures its prosperity. In this view, concrete-surfaced Morgan Street carries the Sunset Highway through town. Visible signs include the Lincoln Hotel, Lincoln Cafe, H. Livingston Hardware, Texaco gasoline, and The Fountain Cafe. (LCHS.)

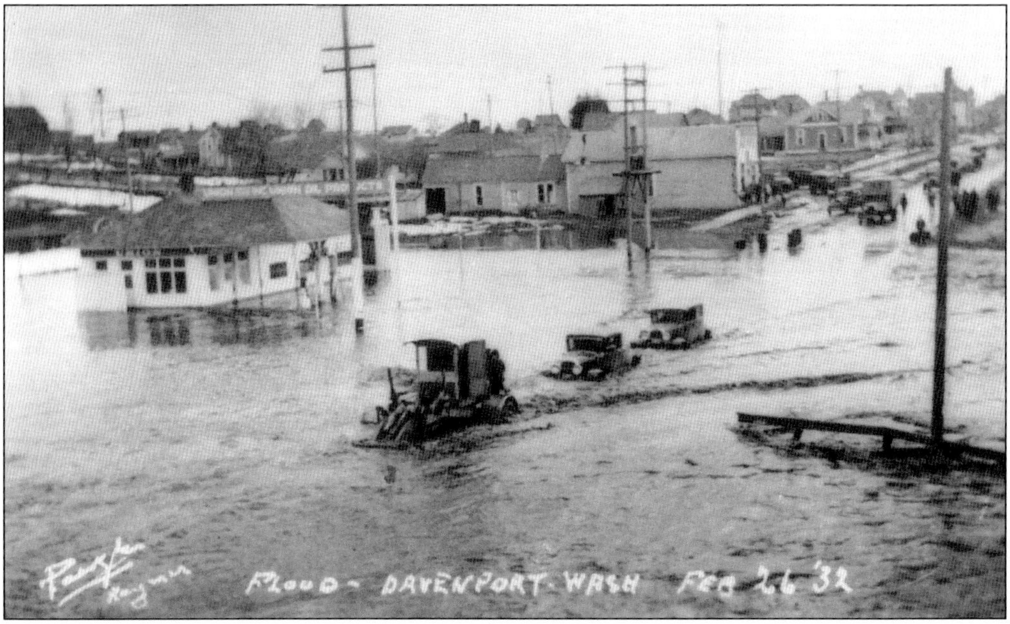

Although Davenport sits amid relatively level farmland, it has often been visited by floods when Cottonwood Creek, which runs through town, overflows its banks. In this view from February 1932, a tractor tows two autos across the flooded intersection of Morgan and 8th Streets past a partially submerged Union Oil Products service station. A truck and additional autos await their turn on higher ground. (LCHM.)

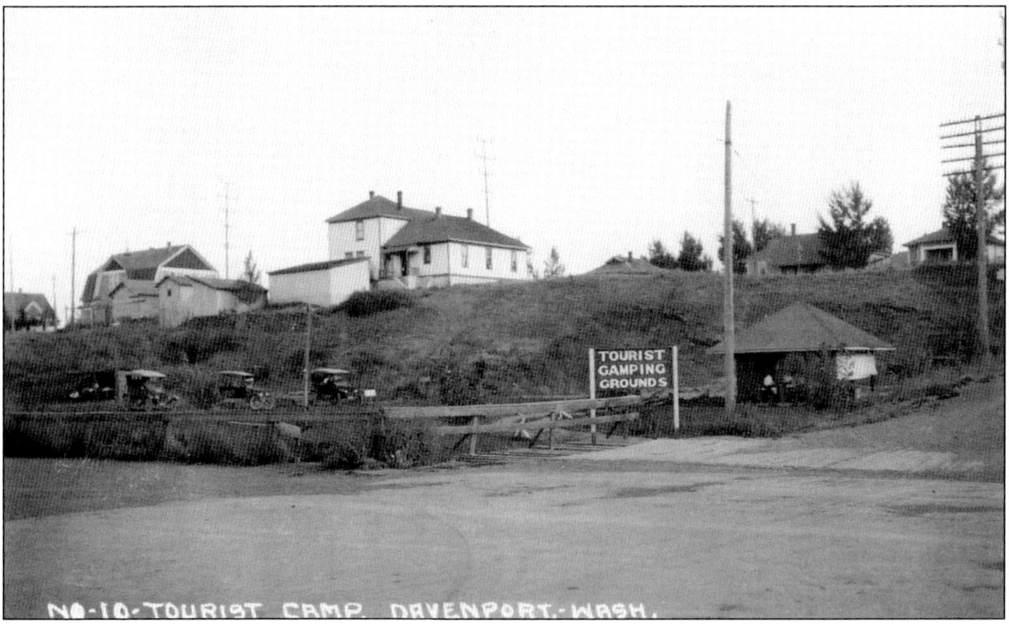

This very early example of a tourist campground was located next to Cottonwood Creek just east of downtown Davenport. The only facilities provided were a communal tent-covered kitchen and picnic tables. Tourists were expected to bring their own lodging and cooking equipment, though restaurants were within easy walking distance. Davenport's city park now occupies the site. (LCHM.)

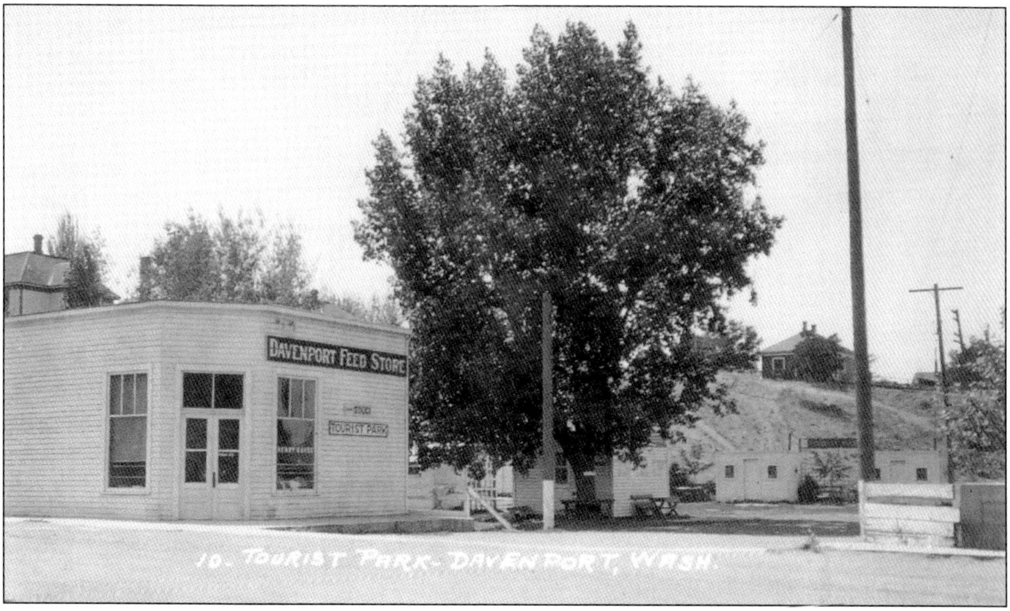

German immigrant Henry Kahse opened a feed store and livery stable at the corner of Park and 7th Streets in Davenport in the early 1900s. By 1930, he had added nine cabins to accommodate the increasing tourist traffic through town. The cabins shared a common kitchen; a separate water closet provided sanitary facilities. Three cabins still exist under massive shade trees opposite the Lincoln County Historical Society. (LCHM.)

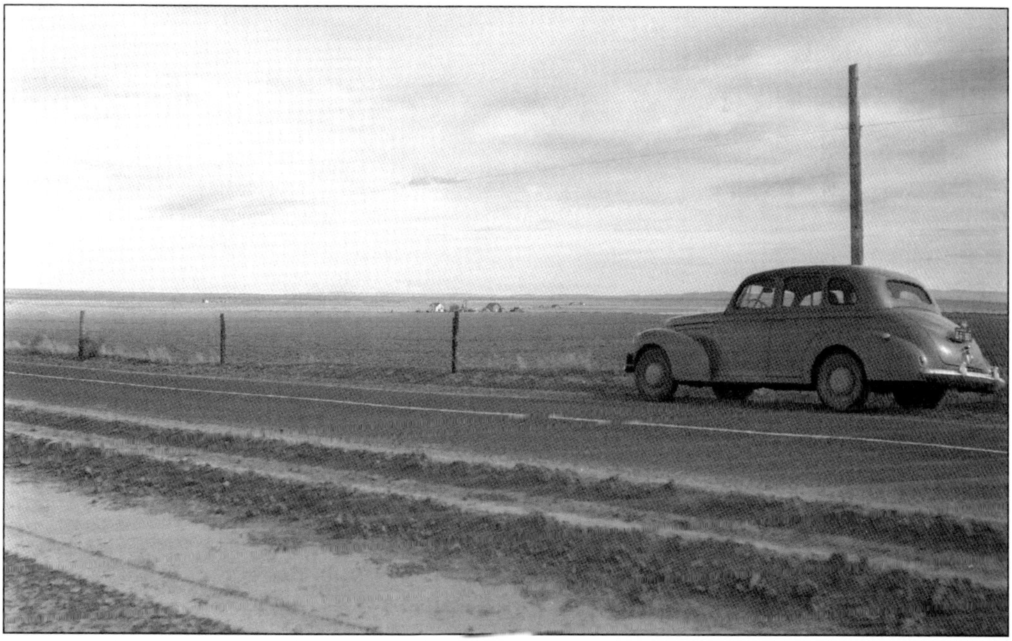

Between Davenport and Quincy, the highway traversed rolling hills and mile after mile of wheatlands. This is wide-open country—trees are rare; farms are scarce, and when they appear in the distance, they often look like miniature towns. In places the fertile farmland is interrupted by scab-rock, evidence of a glacial flood that scoured the area in prehistoric times, scraping the soil down to bedrock. (WSDA, AR-09701002-ph000162.)

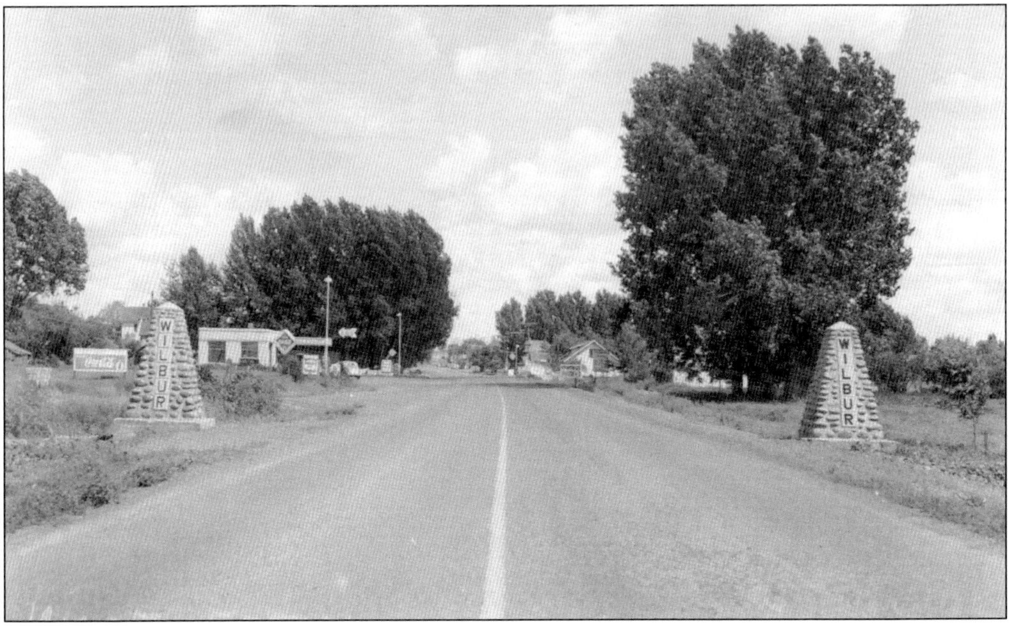

Wilbur boomed during construction of Grand Coulee Dam, 18 miles to the northeast, in the 1930s. Like many towns on the Columbia Plateau, it has lost population since then (881 in 2010, down from 1,011 in 1940) but remains a vibrant marketing town and wheat shipping point. These pylons on the west edge of town once welcomed visitors but have long since disappeared.

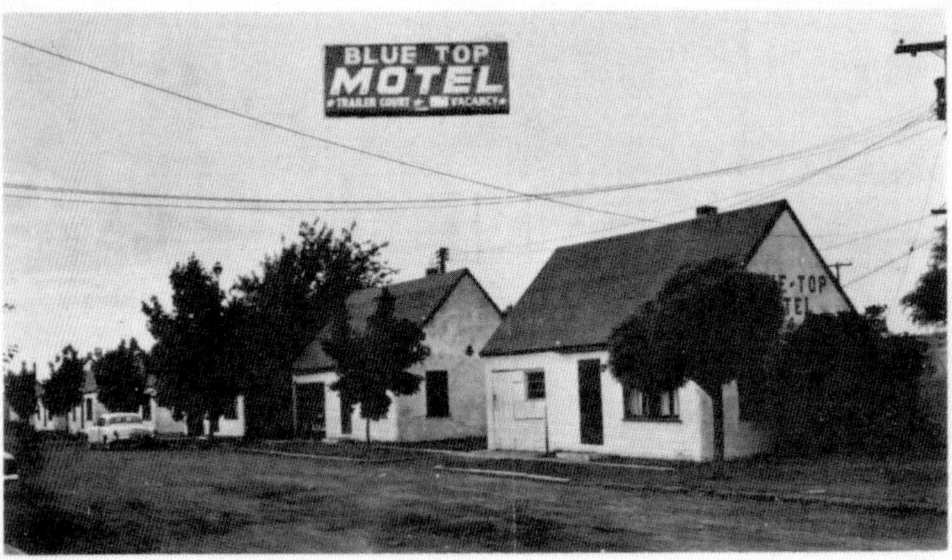

BLUE TOP MOTEL, WILBUR, WASH. On US10, 22 mi. E. of Coulee Dam. Kitchens, Frigid., all-electric

Also at the west edge of Wilbur was the Blue Top Motel, with six air-cooled cottages offering private toilets, bathtubs and showers, and cooking facilities. Each unit had its own garage. Nightly rates in 1948 were $3 for two persons and $5 to $6 for four or more. A trailer court operated in conjunction with the motel. Food and gas could be found a short distance back into town. (JCO'D.)

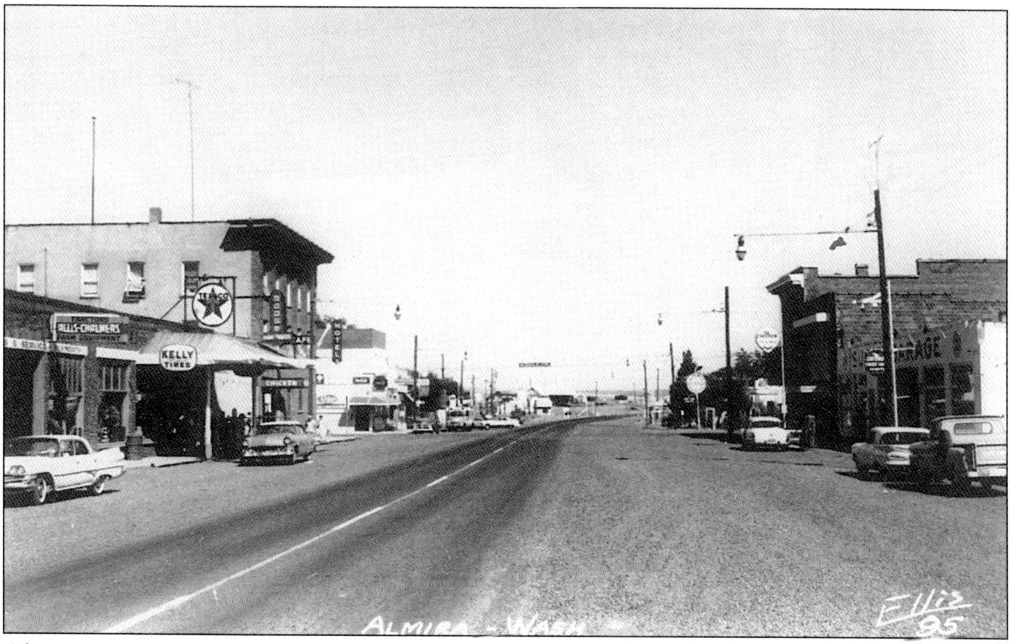

Almira was a small but substantial town astride the highway with a hotel and restaurant, several service stations (Texaco on the left, Chevron on the right with gas pumps at the curb), and a 1939 population of about 500. Today, the figure is about half that number, and Almira's nearly deserted downtown sits a quarter of a mile off a new highway bypass. (JCO'D.)

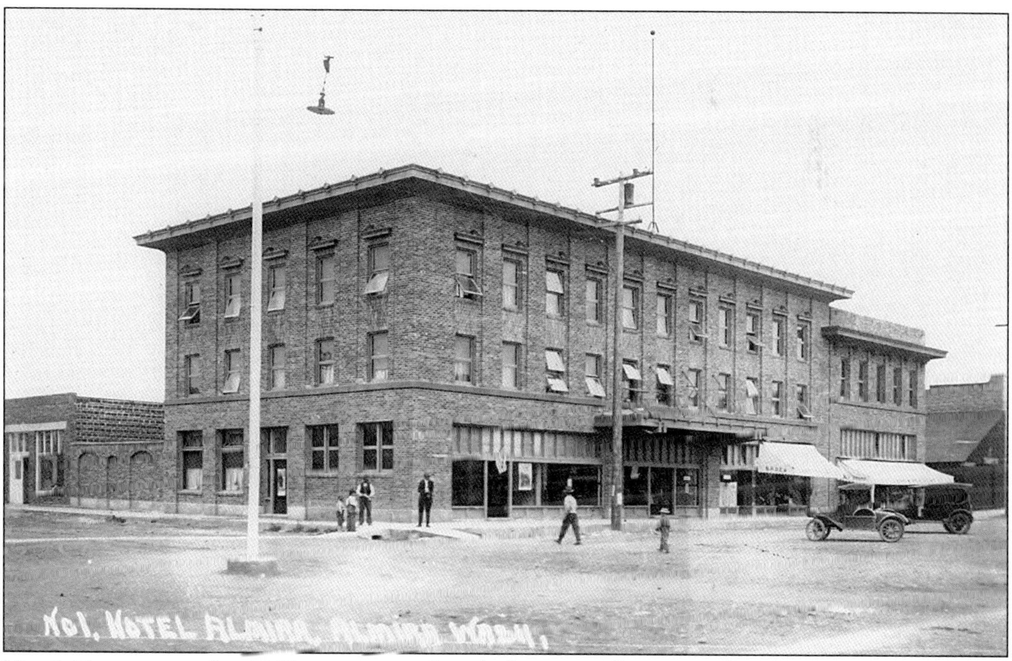

Hotel Almira opened in 1917, just as a new breed of motorists began to take advantage of inexpensive autos and improved roads to see the country after World War I. The hotel advertised itself as "the best equipped hotel between Spokane and Seattle" with over 60 rooms and an on-site restaurant. Closed for years and reputedly haunted, the hotel is currently undergoing restoration. (LCHM.)

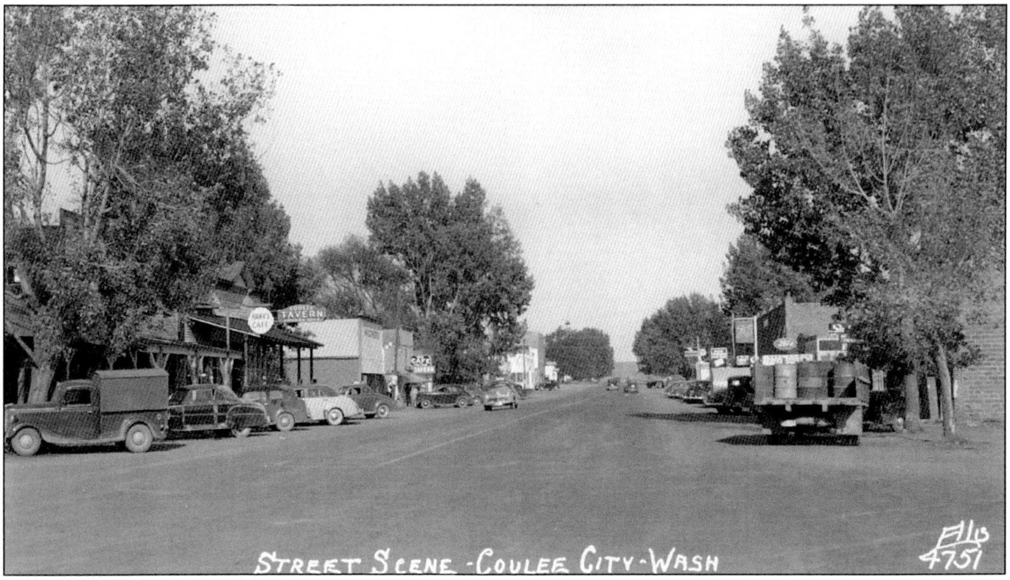

The Sunset Highway passed through Coulee City in early days but later realignments and the construction of Banks Lake left the town about a half mile off the main route. This view from the late 1940s looks east through downtown. Visible here are signs for Hank's Cafe, the Coulee Tavern, the Gem Cafe, and Union 76 and Chevron service stations. The Thompson Hotel is barely visible in the far distance.

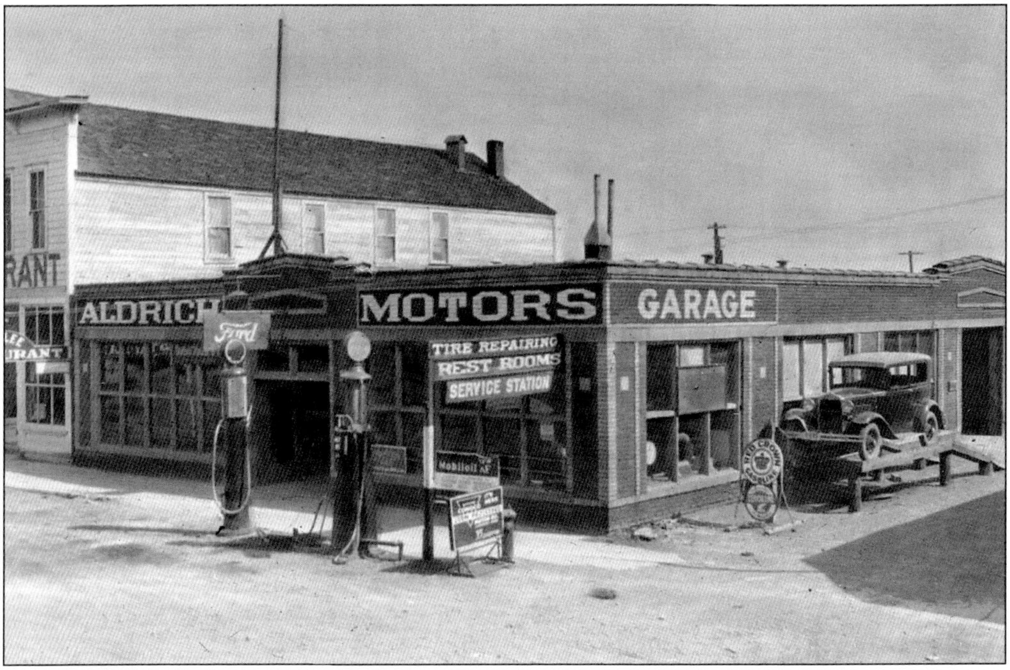

Aldrich Motors in Coulee City, owned by Clifford Aldrich, was a typical early garage with gas pumps right on the street and a makeshift wooden service rack alongside the building. In those days before franchises, Aldrich's offered two brands of gasoline—Red Crown and Conoco—and two brands of motor oil—Mobiloil and Zeroline. The Conoco triangle sign dates the photograph to about 1930. (WSDA, AR-04503146.)

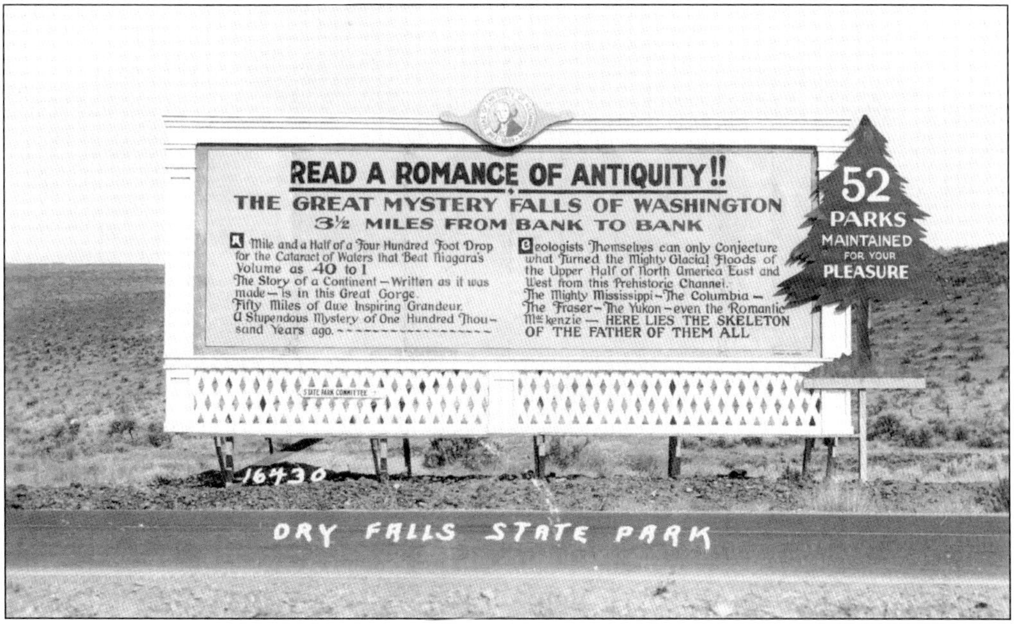

This billboard served up a colorful description to entice travelers to visit Dry Falls, south of the highway a few miles west of Coulee City. The falls are a remnant of the Ice Age, formed nearly 20,000 years ago when an ice dam blocking historic Lake Missoula broke and unleashed a tremendous flow of water over a 400-foot-high, three-mile-wide rock escarpment.

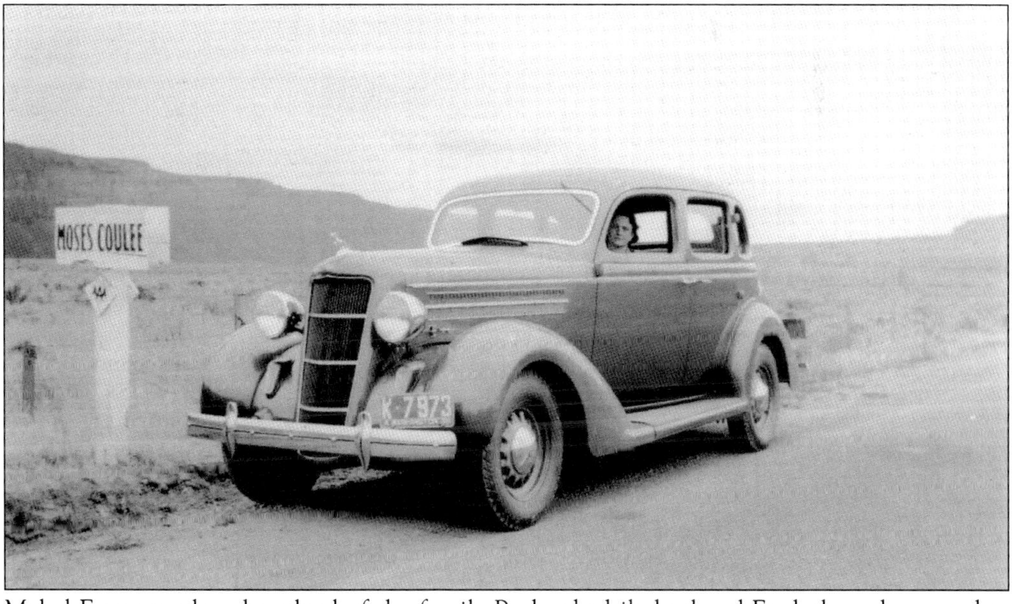

Mabel Fawcett takes the wheel of the family Packard while husband Fred plays photographer in this c. 1935 view taken in the middle of Moses Coulee. While not as spectacular as nearby Grand Coulee, Moses Coulee—a mile wide, 600 feet deep, and extending for over 35 miles to the Columbia River—holds its own for scenic quality.

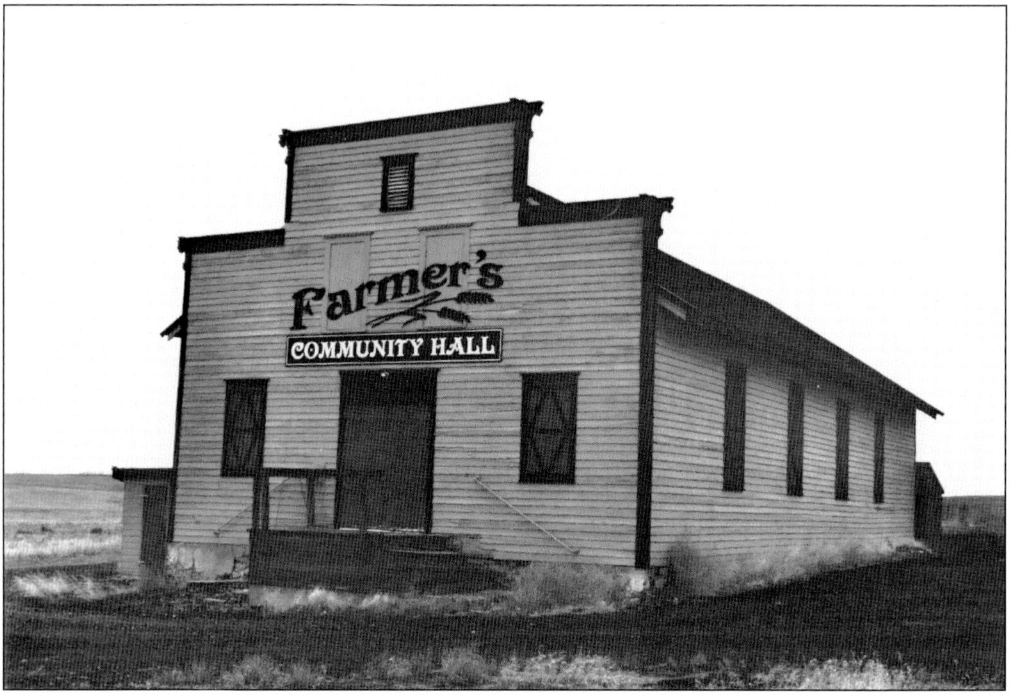

About seven miles east of Moses Coulee, the highway passed through the middle of Farmer. Typical of many small towns scattered across the Columbia plateau, Farmer came into being to serve surrounding farms and ranches at a time when travel, for most people, was limited to about 20 miles from home. Farmer's peak years were between 1905 and 1915 when it had a general store, hotel, restaurant, butcher shop, and blacksmith. Being located on the highway allowed Farmer to endure longer than many other small country towns—as late as 1932 it still had a store. It eventually succumbed as better roads allowed farmers and ranchers to travel to larger cities such as Waterville. Virtually all that remains of Farmer today are these two buildings and a cemetery a quarter mile off the highway.

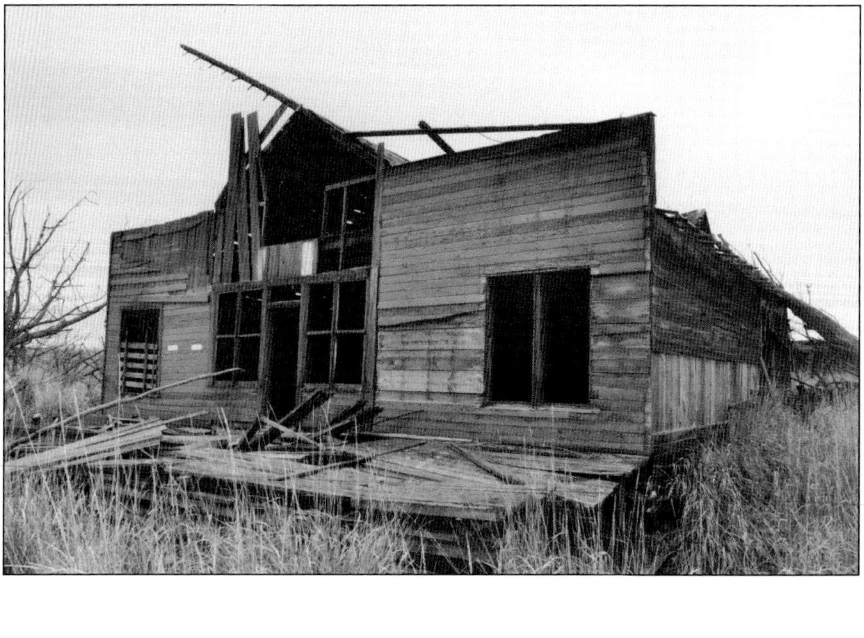

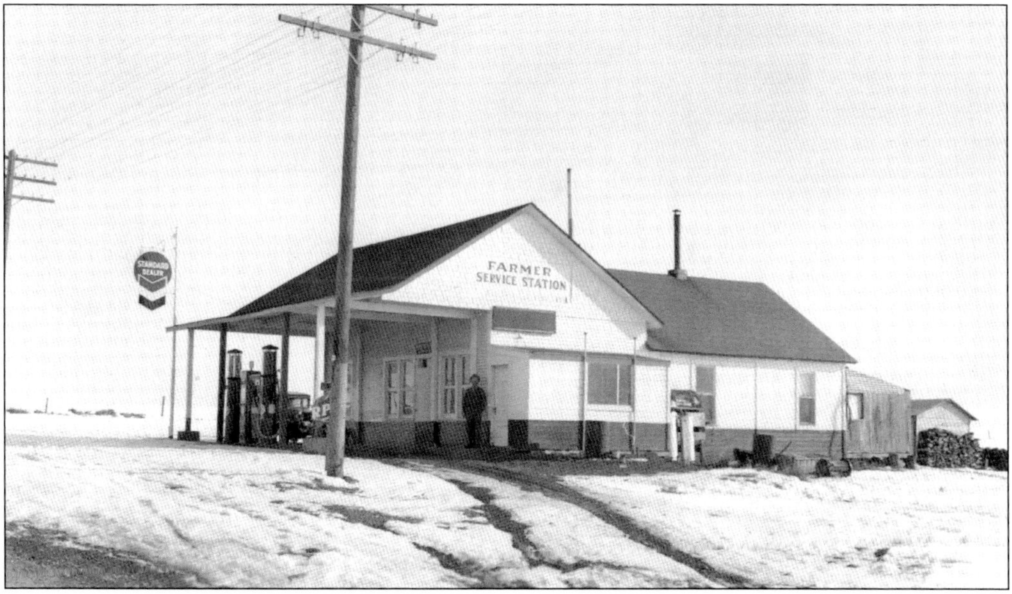

Long after other businesses in town had faded, Farmer continued to support a service station. With two pumps offering Standard brand gasoline, it was the only refueling stop on the 25-mile stretch between Coulee City and Douglas. The station, pictured here in February 1941, survived for several more years but today has vanished, along with the rest of Farmer. (DCHS.)

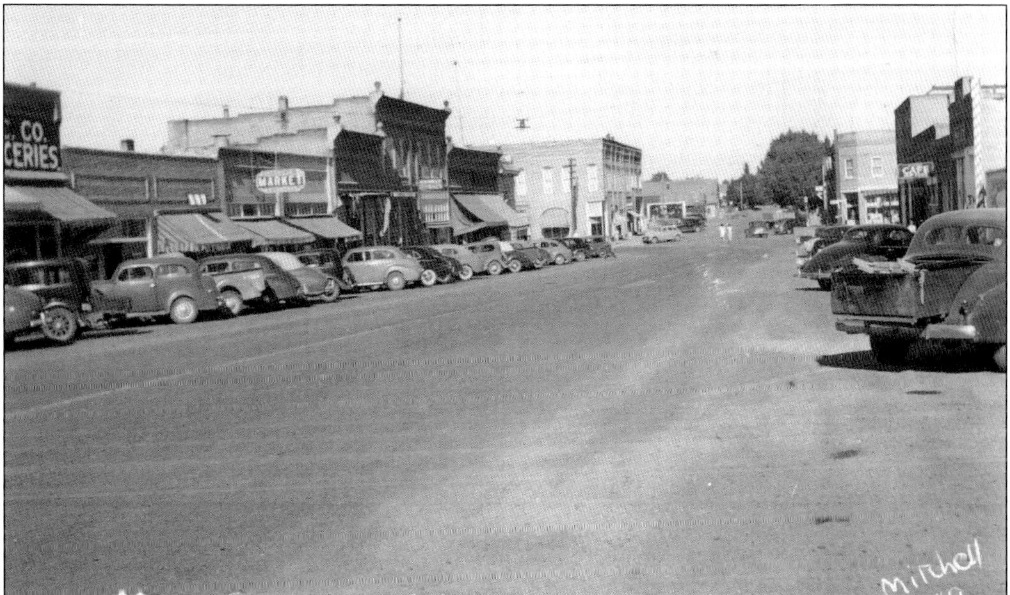

Waterville is the seat of Douglas County. The Sunset Highway approached town from the east via E Ash Street, turned south on N Chelan Street, west on E Locust Street, and finally south again on E Park Street and out of town. This view looks east to the corner of N Chelan and E Locust, the center of town, where many buildings date to soon after incorporation in 1890.

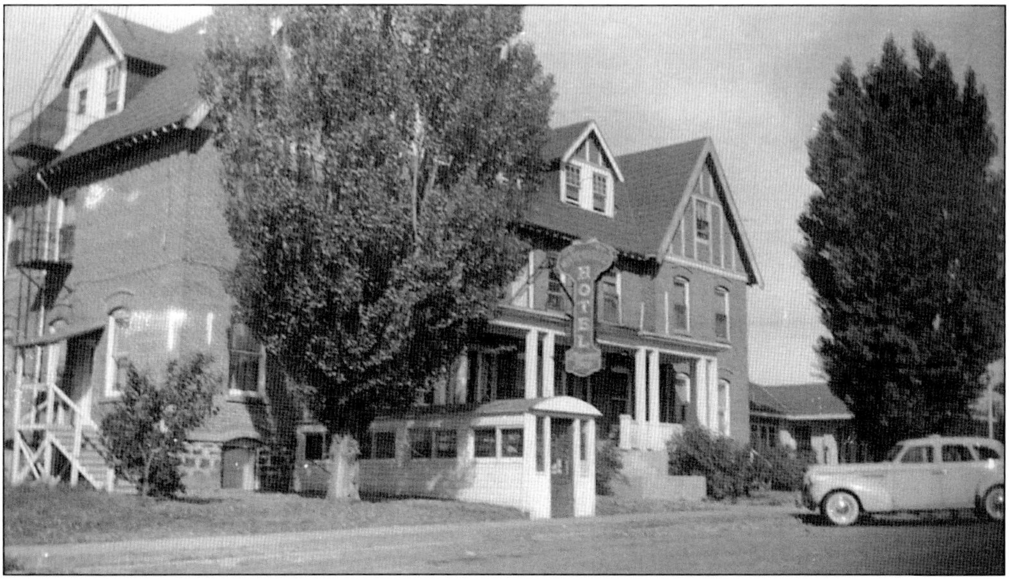

The Hotel Waterville has been serving travelers for over a hundred years. Built in 1903, it is a classic example of a country hotel with high ceilings and claw-foot bathtubs. The hotel anchors the western end of Waterville's small but historic downtown and, at three stories, is the tallest building around. The current owners have restored the hotel to its early elegance. (DCHS.)

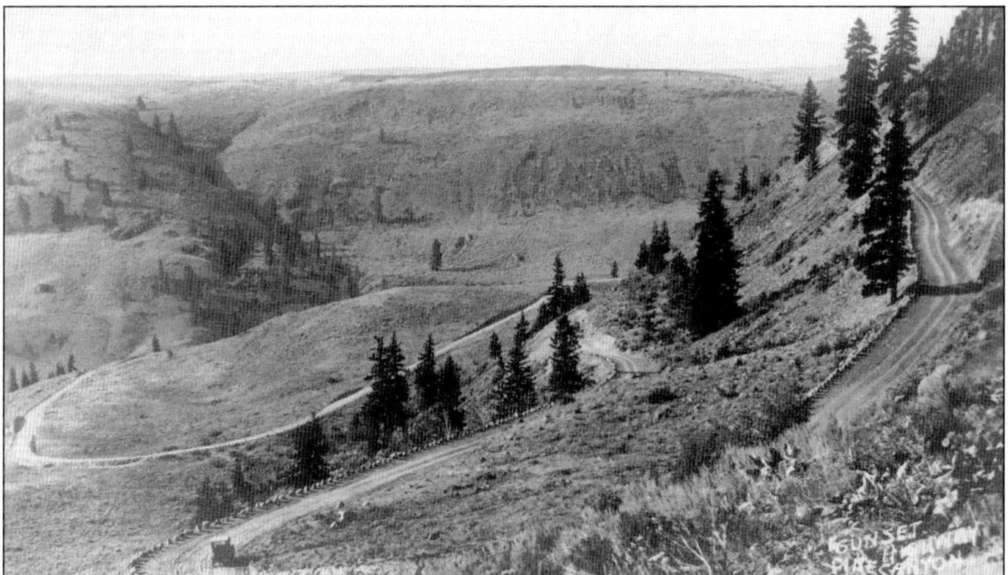

Three miles west of Waterville the Sunset Highway entered Pine Canyon, the start of a long descent to the Columbia River at Orondo—a drop of 2,000 feet in only seven miles. Westbound travelers had gravity to assist the elevation change; for eastbound motorists it was a slow, often hot, uphill struggle to make the grade. US Highway 2 still traverses the canyon, though on a straighter alignment.

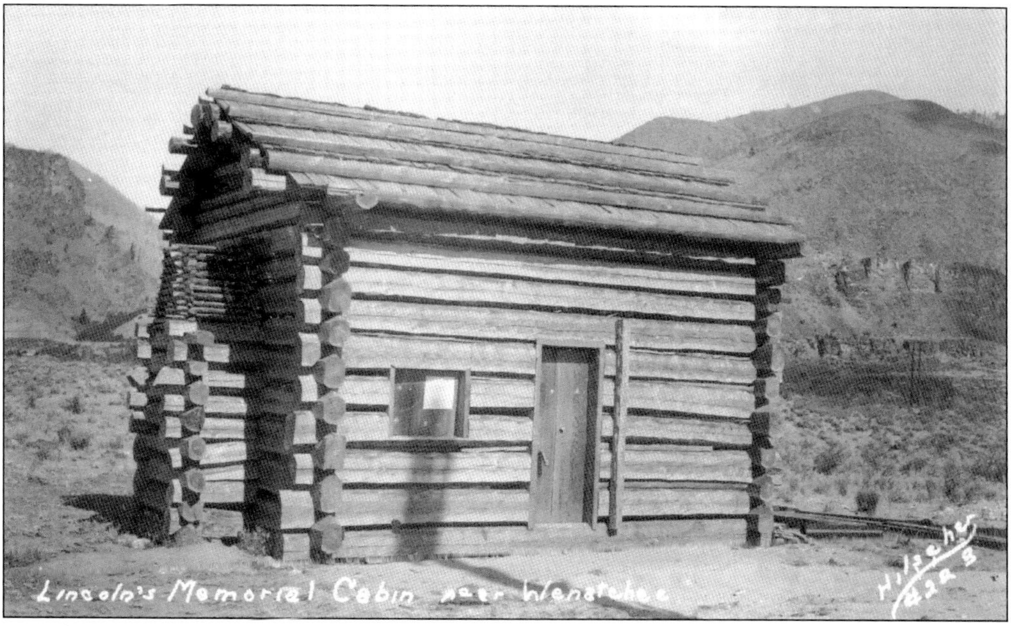

Lincoln Rock is a geological oddity resembling the profile of Abraham Lincoln. The rock sits above the west bank of the Columbia River but can only be properly viewed from a point on the east bank about 15 miles south of Orondo. This little cabin stood along the Sunset Highway to mark the viewpoint and remind visitors of Honest Abe's log cabin heritage. (DCHS.)

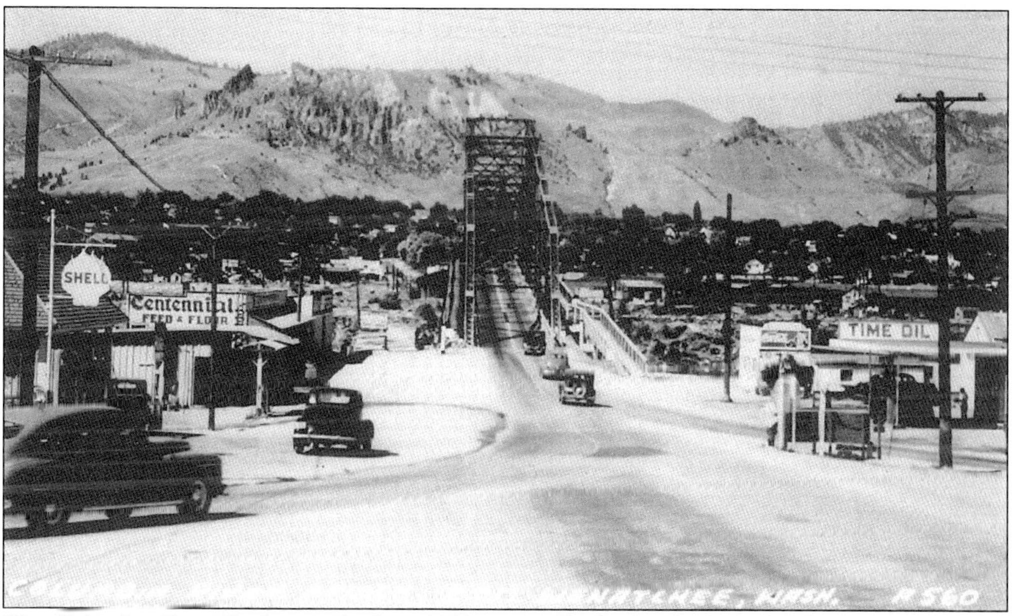

The Columbia River Bridge was built in 1908, reputedly the first highway bridge to cross the river south of Canada. In this view, the Sunset Highway comes in from the right (north) and turns past Time Oil and Shell service stations to cross into Wenatchee. Locally known as Pipeline Bridge for the irrigation pipeline it carries, the old bridge was retired in 1950 and now conveys foot traffic. (WVM, 008-14-1.)

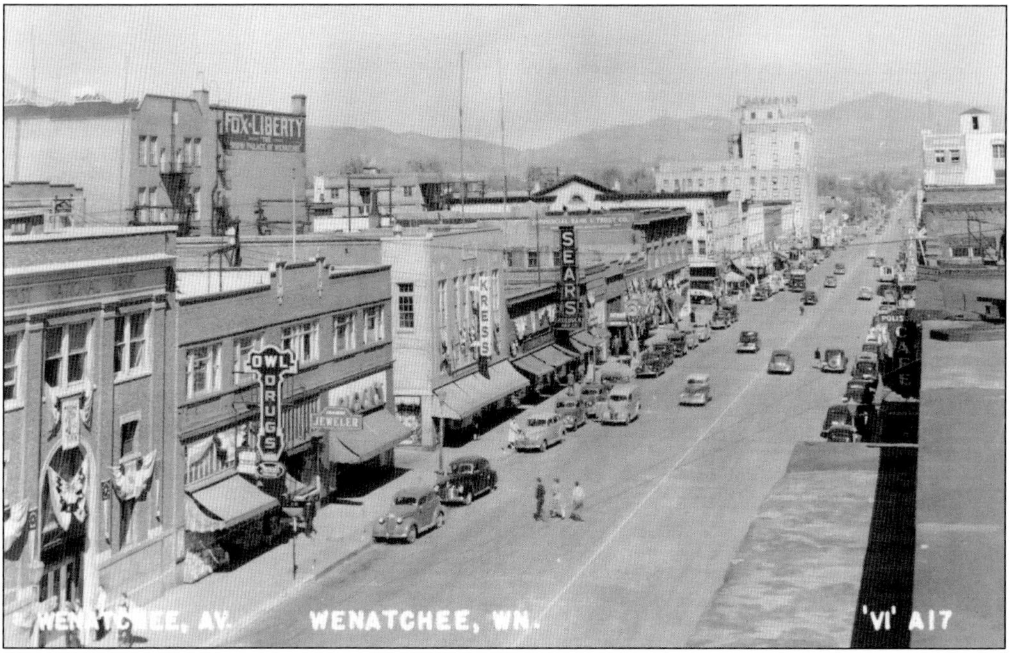

Downtown Wenatchee was a busy place, as this late 1940s view shows. Among the many businesses whose signs can be seen along Wenatchee Avenue, the route of the Sunset Highway through the city, are Owl Drugs, Kress' and Sears Roebuck's department stores, the Olympia Hotel, the Rialto Theatre, and the Cascadian Hotel. A street clock towers over autos parked near Sears, with two more clocks in the distance.

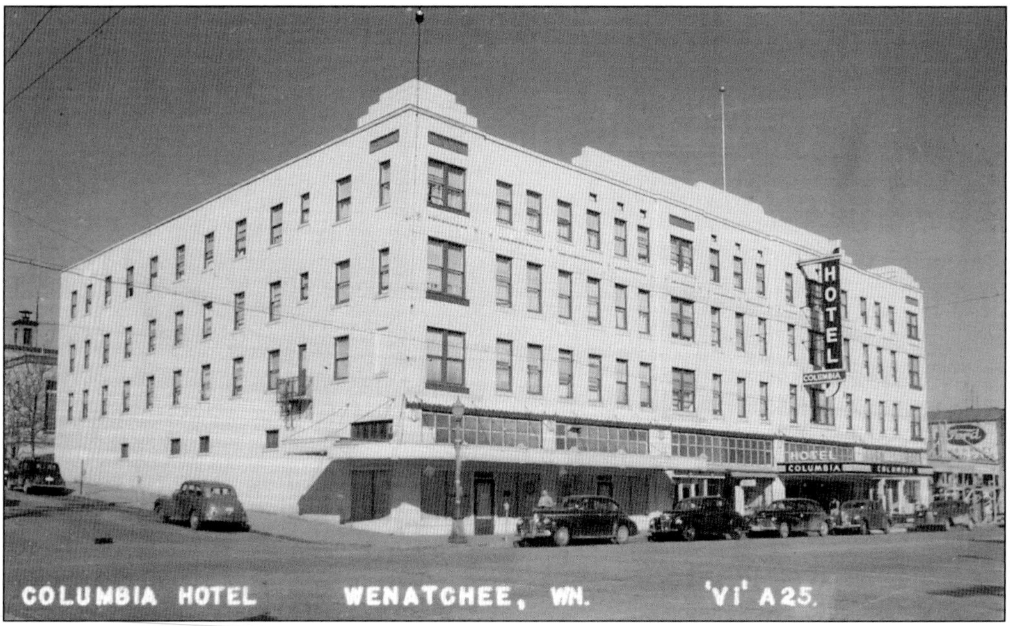

The four-story, 200-room Columbia Hotel described itself as "one of the outstanding hotels of Eastern Washington." Room rates started at $1.50 a night, with a dining room (with "Distinctively Good Food Served"), a coffee shop, and a fountain on the premises. Located in downtown Wenatchee at 130 Wenatchee Avenue S, the site is now a park. (JCO'D.)

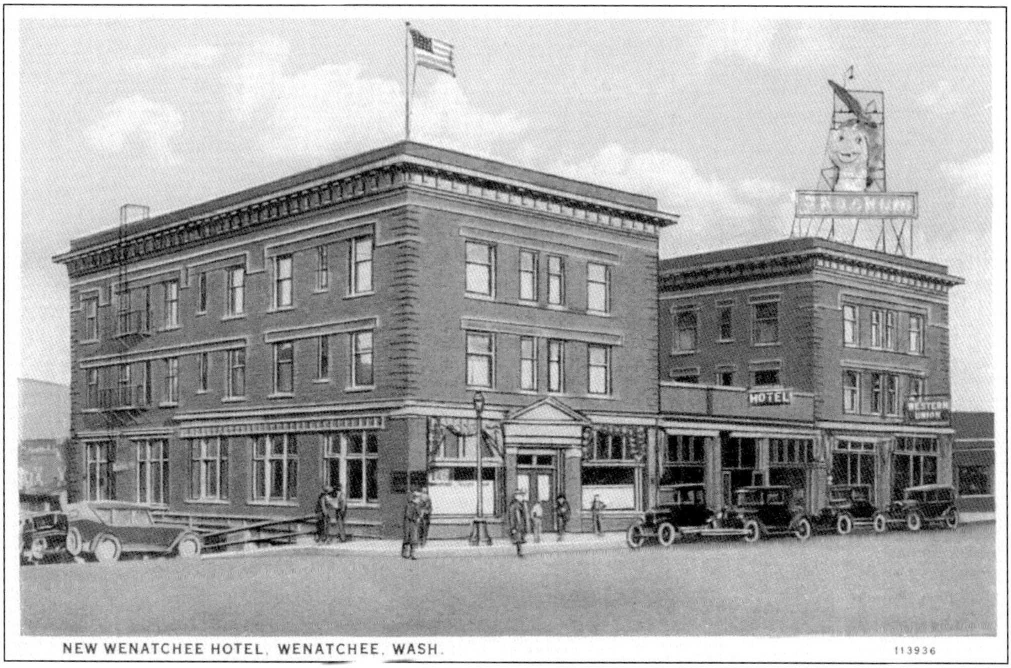

On the opposite end of the block from the Columbia, the New Wenatchee Hotel was an older, more genteel establishment with 68 rooms at $1.25 and up ($1.50 and up with shower bath). A café was located within the hotel. The Skookum sign atop the hotel dates to about 1913. Originally an advertisement for the Skookum fruit cooperative, the sign was restored and relocated atop an Office Depot on N Wenatchee Avenue.

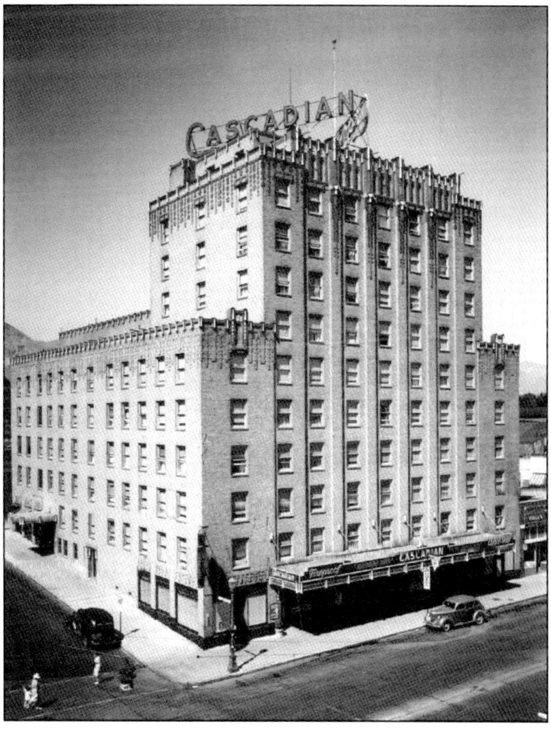

The Cascadian was the crown jewel of Wenatchee's hotels—10 stories of Art Deco styling, built in 1929 and still the city's tallest building. Completely modern and fireproof, the Cascadian had 200 rooms, air conditioning throughout, banquet and meeting rooms, and a radio station broadcasting from the top floor. The building, at 102 N Wenatchee Avenue, has been converted to apartments. (WVM, 75-23-44.)

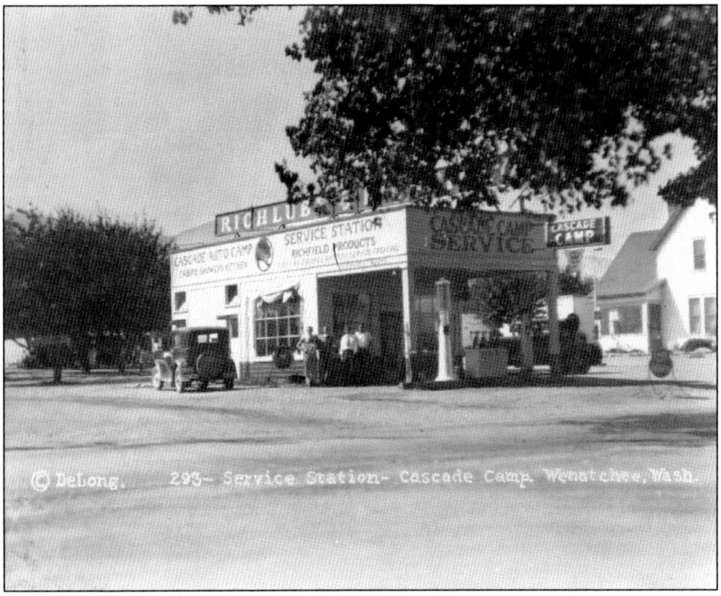

In 1938, Maud M. Cain was operating the Cascade Auto Camp at 902 N Wenatchee Avenue. The 27 "modern cabins with modern equipment" were available at a nightly rate of $1.25–$2.50. The units were arranged in a U shape with this service station—festooned with signs for the camp and Richfield gas—in the middle. The camp was still in business in 1950 as Cascade Court. (WSDA, AR-04503192.)

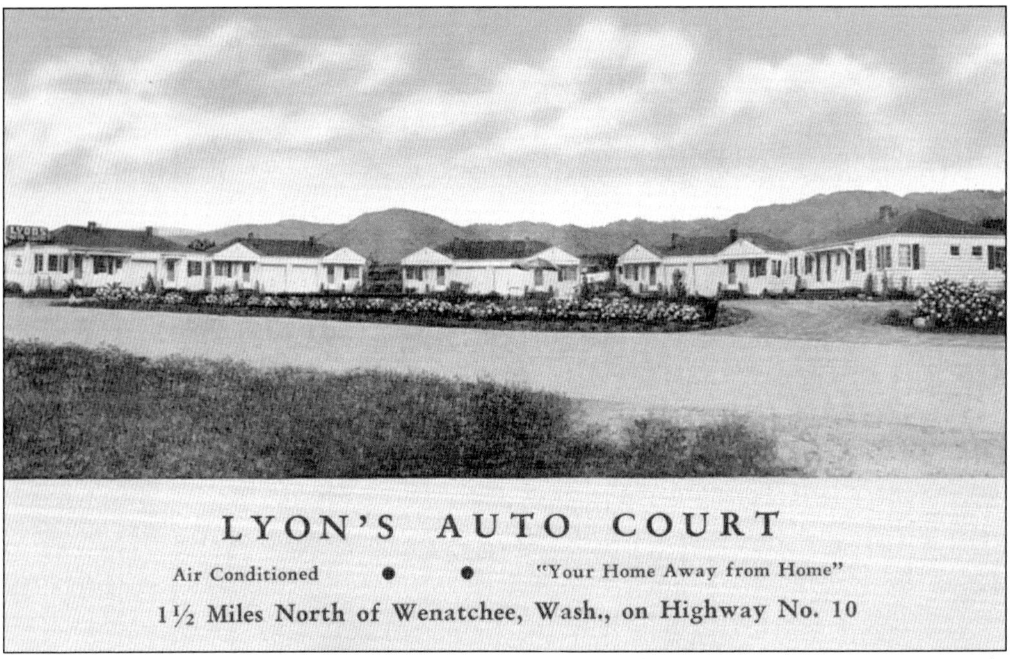

Lyon's Auto Court was at a crossroads on the north edge of Wenatchee. The 37 cottage-style units arranged in a semicircle fashion faced a greensward of trees and flowering shrubs. In 1961, the Lyons purchased another motel—the Starlite, just up the highway—and gave it to their son and daughter and her husband to start them in the motel business. Lyon's is gone; the Starlite is now a Value Inn.

The Windmill, built in 1931 as a roadhouse at the edge of Wenatchee, acquired a shady reputation in its early years. It was sold several times over the following decades, each new owner elevating its stature as a steak house—nearly 360,000 steaks were served between 1962 and 1982. The present owners have preserved the comfy booths and knotty-pine interior as well as the Windmill's high standard of service.

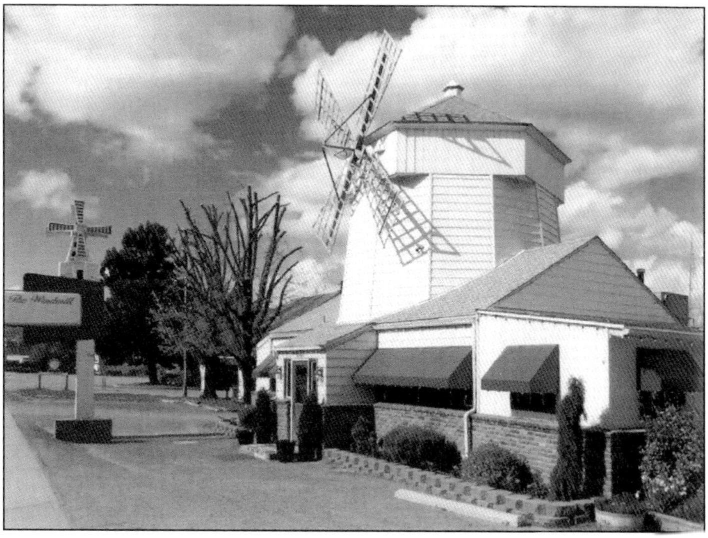

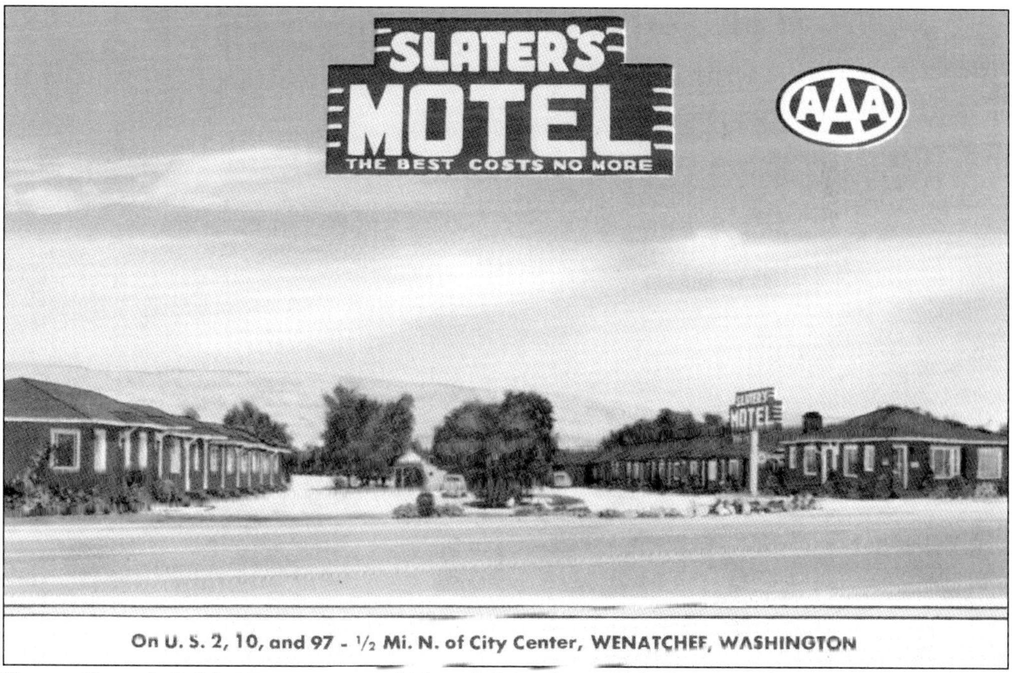

Emory C. and Golda Slater operated Slater's Motel at 1505 N Wenatchee Avenue next door to the Windmill. Rooms featured air conditioning, central heat, tiled showers, kitchens, and Beautyrest mattresses. Bill and Mildred Talbott purchased the motel in July 1959, selling it in 1962 to a group of local businessmen. Under management of Mr. and Mrs. Fred Maydole, it was renamed the Bel-Air Motel. It is no longer in business.

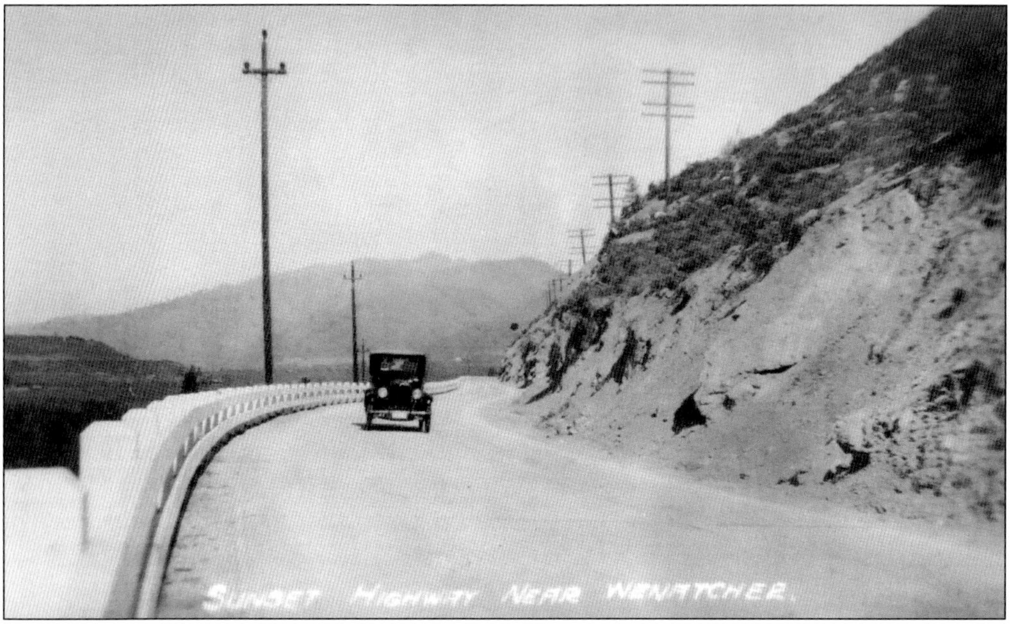

Motorists had smooth highway ahead of them after they left the city and proceeded up the valley of the Wenatchee River. The road was good, the many orchards through which it passed made for a delightful journey, and traffic was light—so light that this driver occupied the center of the highway fearless of oncoming vehicles. This view looks west, up the valley.

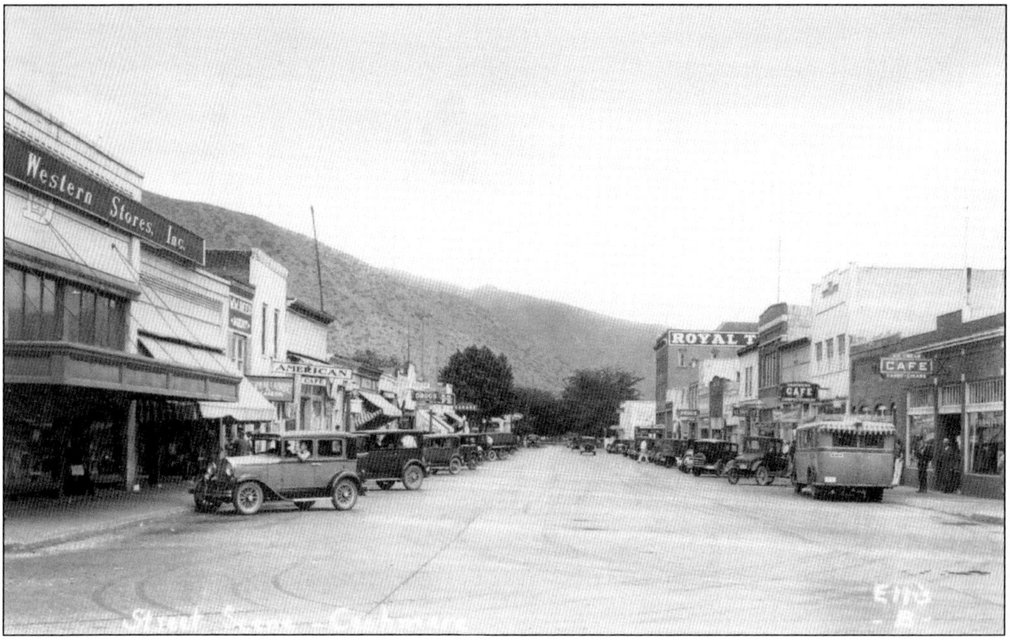

Cashmere, 12 miles west of Wenatchee, was first called Mission and then Kashmir, later Americanized to the present spelling. Cashmere boomed when irrigation turned dry valley lands into lush orchards, and continues to be a fruit-packing center. The largest town until Cle Elum, nearly 50 miles away, Cashmere offered full services to travelers: two garages and three cafés—Crosley's, the Cashmere, and the American can be seen here.

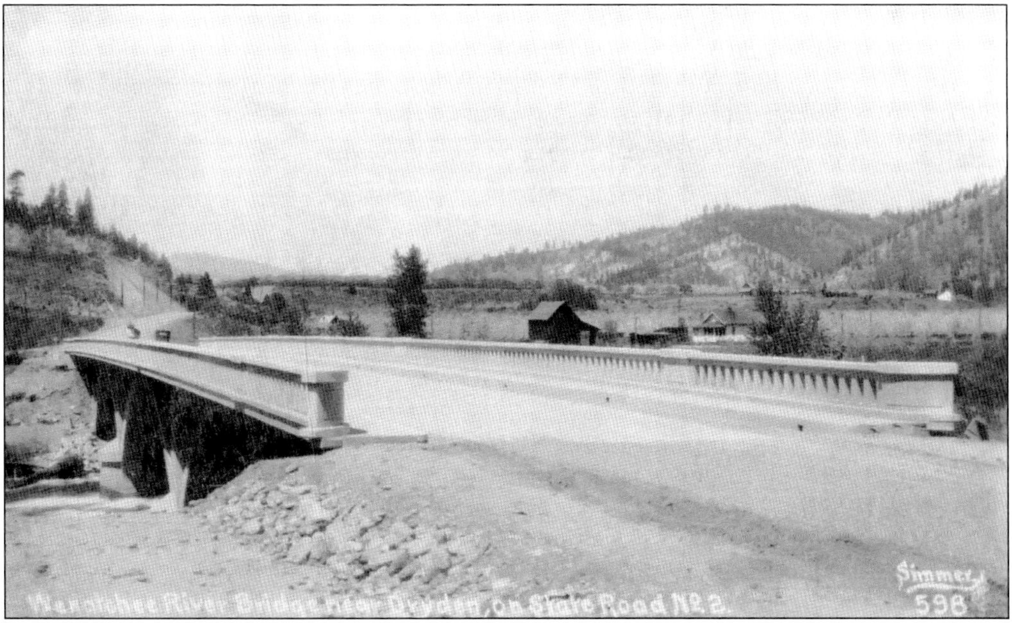

A recently completed bridge—so new that the highway leading to it has yet to be paved—crosses the Wenatchee River half a mile west of Dryden in this 1920s photo. Dryden sits in apple-orchard country. Main Street was once the route of the Sunset Highway through town, but realignment bypassed Dryden in the 1970s and replaced the graceful bridge shown here. (WVM, 75-23-16.)

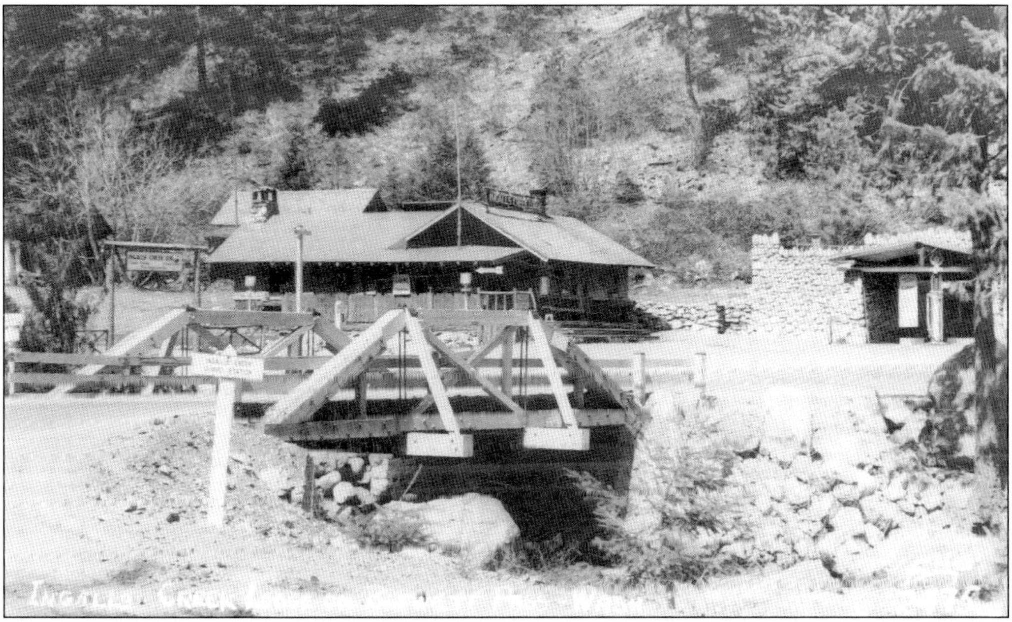

Ingalls Creek Lodge was located about six miles southwest of Dryden on a now-abandoned early alignment of the Sunset Highway. The lodge offered cabins, a restaurant, a store, and a service station and was adjacent to Ingalls Creek, which formed a waterfall at this point, and the Ingalls Creek Forest Camp of the Wenatchee National Forest. The lodge later moved to its current location on Highway 97.

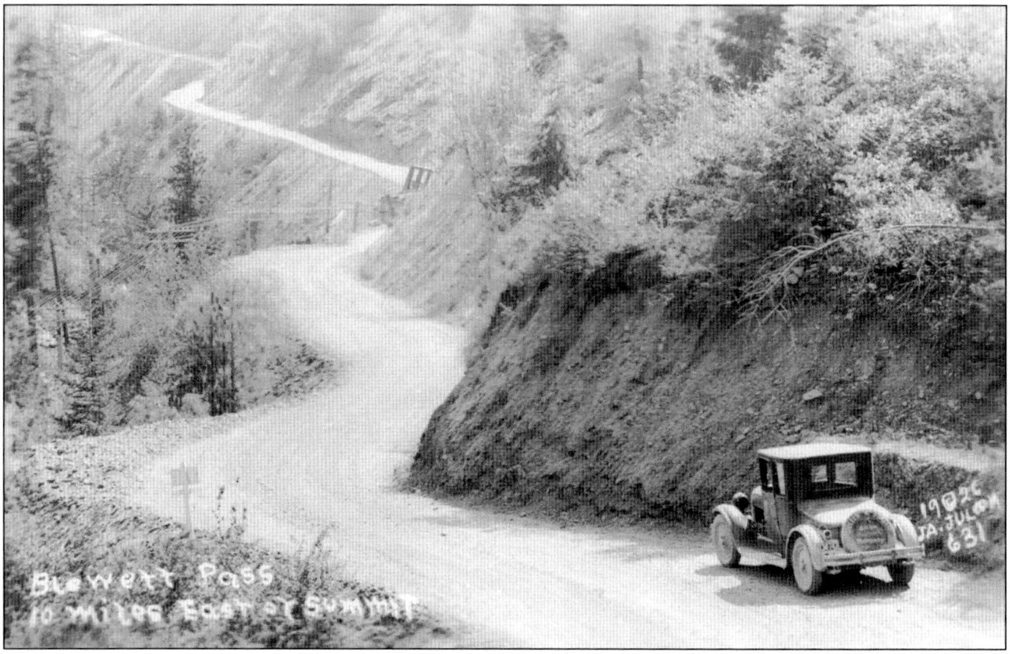

The climb to Blewett Pass from the north was a long and winding road, as this 1926 view shows. Up the hill from the auto, peeking around a bend in the road, an old piece of earth-moving equipment is visible—not surprising since this section of highway required nearly constant maintenance. The hillside, once denuded of trees, has since regrown with a thick pine forest.

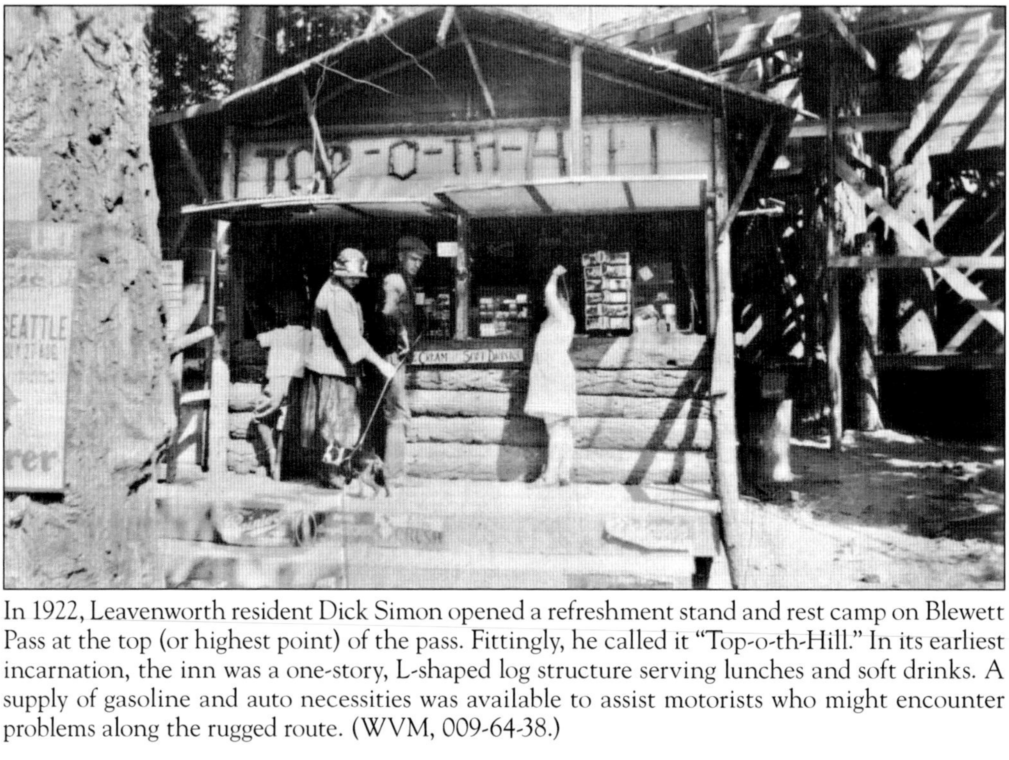

In 1922, Leavenworth resident Dick Simon opened a refreshment stand and rest camp on Blewett Pass at the top (or highest point) of the pass. Fittingly, he called it "Top-o-th-Hill." In its earliest incarnation, the inn was a one-story, L-shaped log structure serving lunches and soft drinks. A supply of gasoline and auto necessities was available to assist motorists who might encounter problems along the rugged route. (WVM, 009-64-38.)

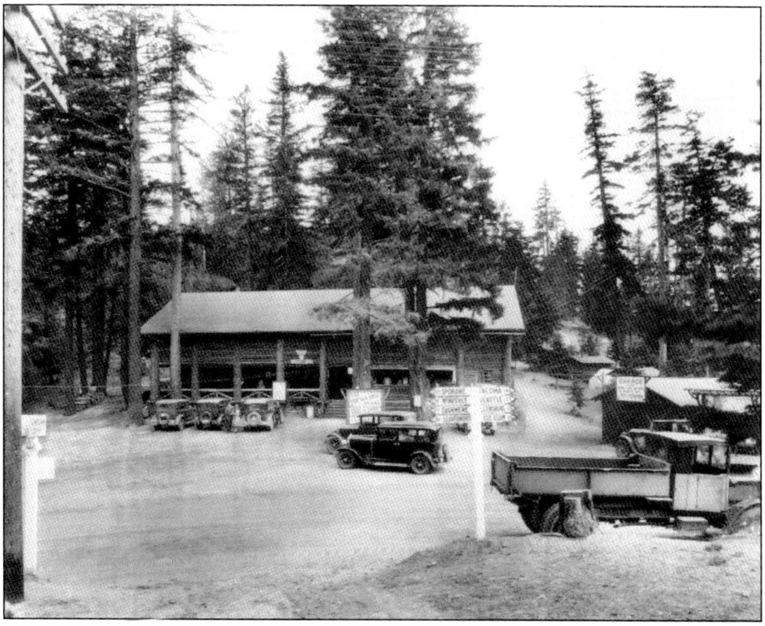

By the late 1930s, the Top-o-th-Hill Inn (also variously referred to as a lodge or tavern) had become a massive two-story mountain lodge with nine tall log entrance pillars. The inn was renowned for its dinners (especially locally caught trout) and offered short orders for those in a hurry. Adjoining the main building was the Top-o-th-Hill Service Station. (EverettPL, J1878.)

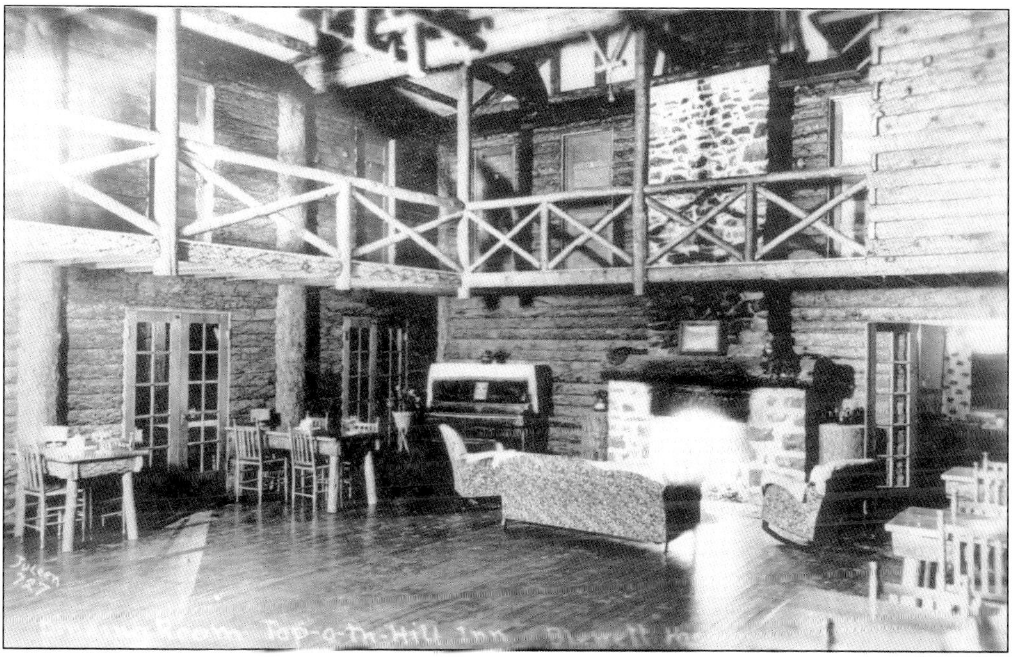

The hunting lodge motif continued inside with a massive stone fireplace, unpeeled logs, exposed beams, tables and chairs of natural wood, and comfortable overstuffed sofas. Modern accommodations were available both in the lodge and in detached cabins scattered on the hillside behind the main building. Rerouting of the highway over Swauk Pass deprived the inn of its business; it has long since disappeared.

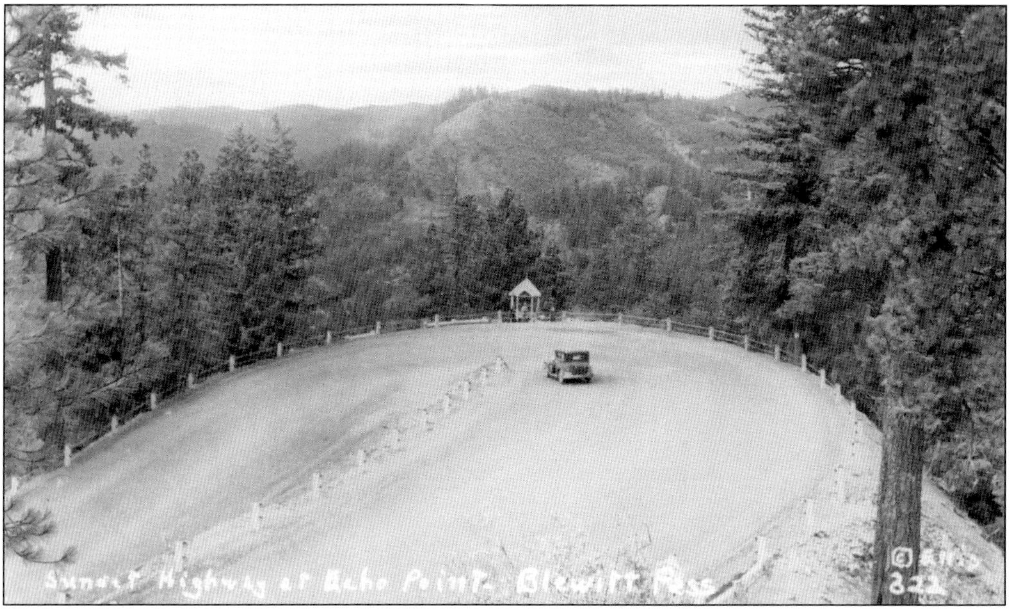

The Sunset Highway over Blewett Pass was famous for its scenery. A pergola-style viewing stand was located at Echo Point south of the summit in early days, allowing travelers a lookout onto the surrounding mountain scenery. The highway was (and still is) very curvy and was abandoned—though still drivable—when US Highway 97 was constructed five miles to the east in the 1950s.

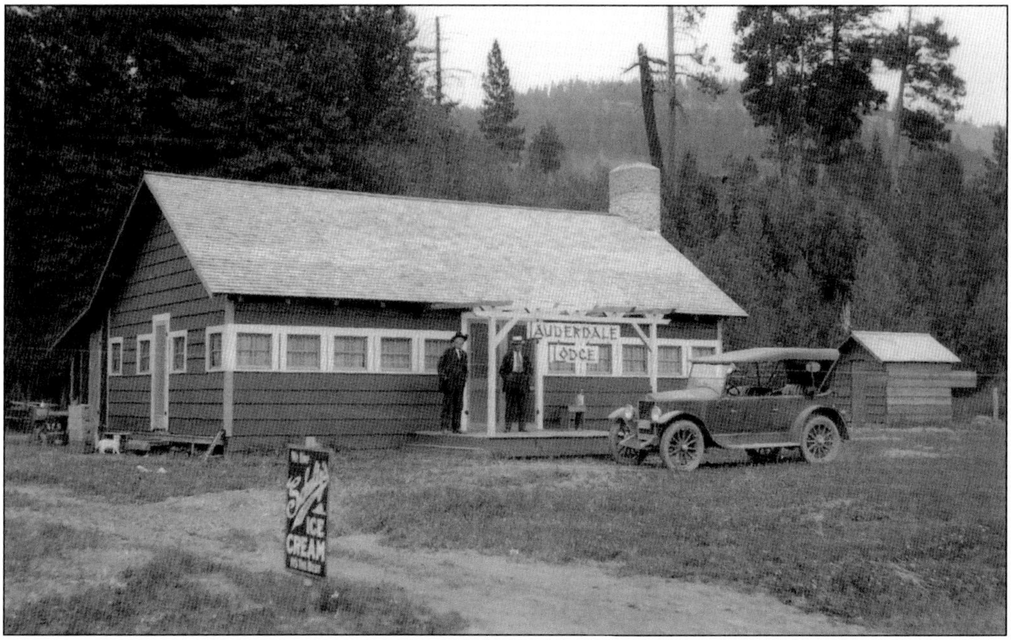

In 1921, H. Lloyd Lauderdale built the Lauderdale Lodge and adjoining service station at Virden, where a road to Ellensburg forked off from the Sunset Highway. The lodge stayed put, but the service station relocated in 1922 and again in 1928 as highway alignments changed. Lloyd's father Frank and son-in-law Ed Snell managed the lodge and station from 1925 to 1933; by then, Virden was known as Lauderdale. (EllensburgPL, BBS338.)

Three

Spokane to Cle Elum via Vantage and Ellensburg

The year 1939 found the Blewett Pass Highway still designated the official alignment of Highway 10, but as early as the mid-1920s, cross-state motorists were seeking alternatives to that route, which was steep and slow-going at best and often closed by snow in winter.

A route southwest from Spokane, crossing the river somewhere south of Wenatchee, would avoid the seasonal closing of Blewett Pass. On paper, the route would follow the existing Central Washington Highway from Spokane through Sprague to Ritzville, then head west to a crossing of the Columbia River east of Ellensburg.

There were problems. Only one substantial town existed between Ritzville and Ellensburg—Moses Lake (called Neppel in those days). Services for travelers would be nearly nonexistent over most of the 120-mile stretch. Another impediment was the river itself. For 50 miles south of Wenatchee, the river is bounded on both sides by bluffs rising as high as 2,000 feet above water level. No obvious easy passage revealed itself.

Highway planners persisted, however. As early as 1915, a ferry was crossing the Columbia River at Vantage; from there traffic could follow a shallow canyon west towards Ellensburg. A new road was proposed to conquer the bluffs east of the river and funnel traffic to the Vantage crossing via Quincy. When a bridge was constructed at Vantage in 1927, State Road 7 was open for travel.

In the late 1940s, a new alignment was built from Ritzville to the recently founded town of George; by 1950, the Vantage Bridge route was officially Highway 10. Additional realignments occurred over the years as bypasses at Ritzville and Moses Lake channeled motorists around, rather than through, towns. Construction of Interstate 90 relegated old Highway 10 to the status of a frontage road across most of this segment; with hardly any traffic using it today, the old road is a pleasant alternative to the crowded freeway.

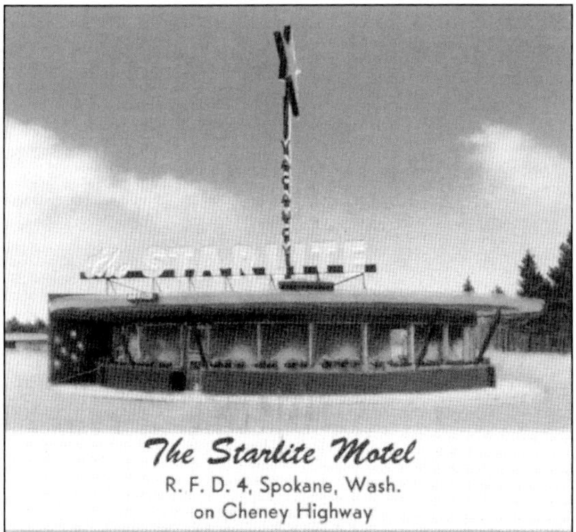

Seven miles west of Spokane, the rerouted Highway 10 branched south from the original Sunset Highway towards Cheney. The "ultra-modern" Starlite Motel was located two miles south of the junction. The motel had 62 units, each with a kitchen complete with cooking utensils, a heated pool, and free television. Its location opposite Geiger Field (now called Spokane International Airport) appealed both to air and highway travelers.

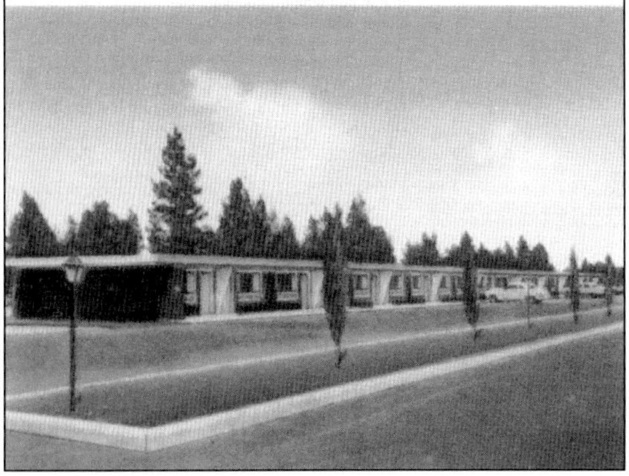

As home of the Eastern Washington College of Education (now Eastern Washington University), Cheney was, and is, a busy college town. It also benefitted from the 1940s rerouting of the Sunset Highway through downtown. Signs for Ben Franklin, a Malt Shop, and Riley's Tavern can be seen in this view along First Street, which appears to be missing center-line stripes.

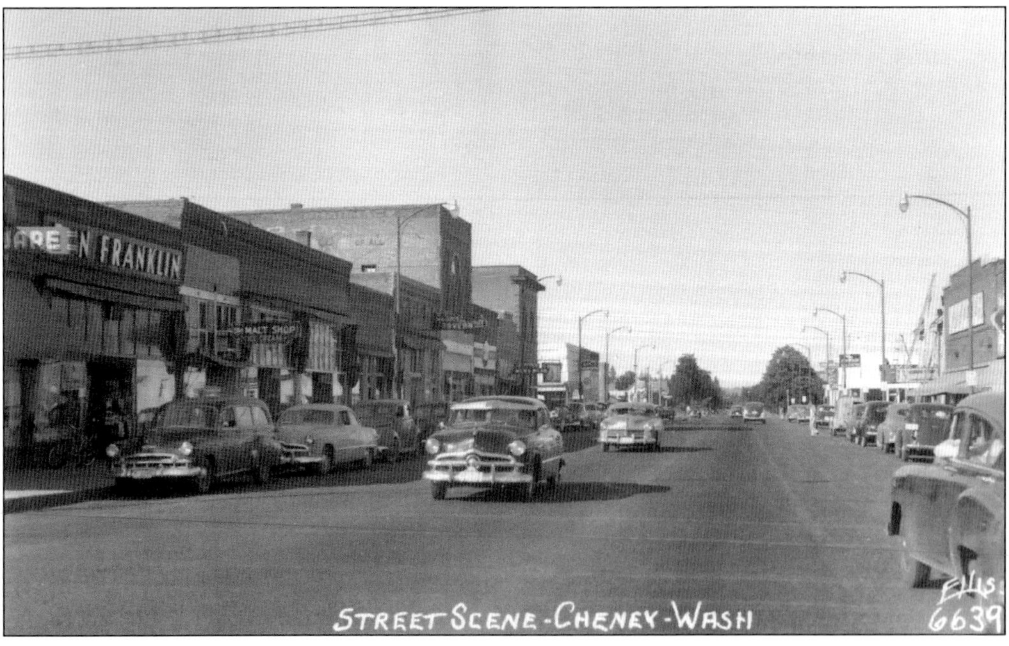

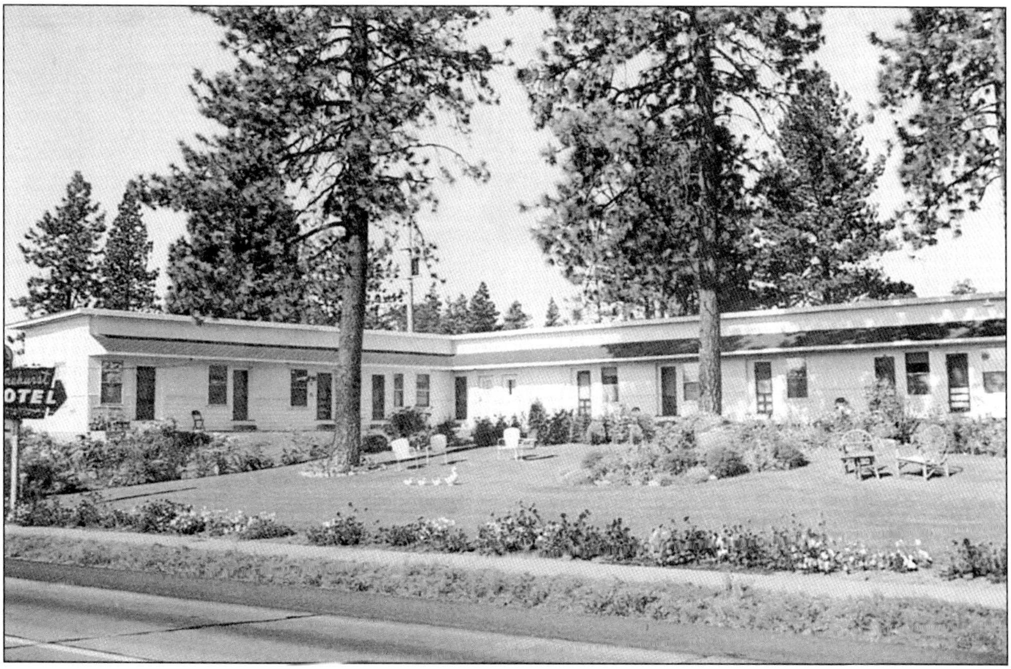

The Pinehurst Motel, under the ownership and management of Tom and Lois Matney, was one of three motels in Cheney in 1953. The Pinehurst was located at the eastern edge of town. Its nine "ultra-modern units," three of which were equipped for housekeeping, offered private baths with showers, individual electric heat, and Deep Sleep mattresses. The building was still standing but vacant in 2010.

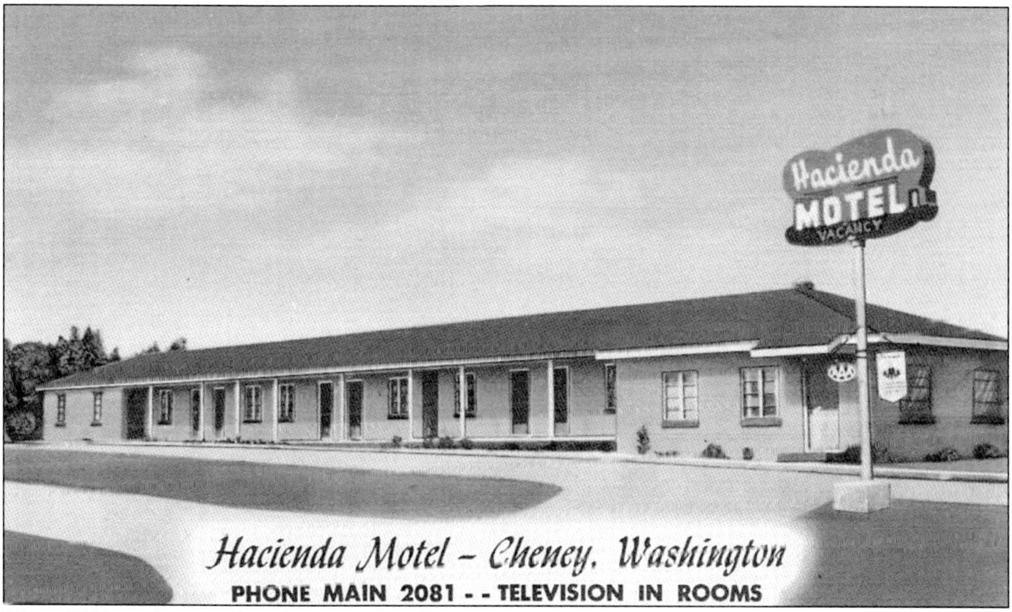

Harry F. Petrick owned and operated Rollie's Cafe in Cheney from 1944 until 1948, when he built the Hacienda Motel. Located five blocks west of downtown, the motel's 10 units featured Simmons Beautyrest beds, refrigerators, and electric heat and was AAA-approved. Petrick, who was also Cheney's chief of police, continued to operate the motel until moving to Spokane in 1960.

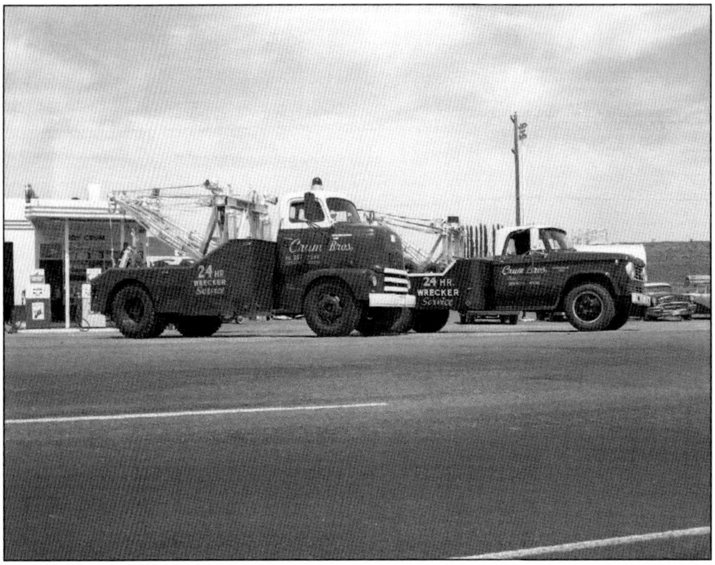

Crum Brothers' wrecker trucks stand ready to assist motorists in this c. 1960 photo. The service station was located at the west end of Sprague and still stands, though no longer pumping gas or providing tow service. Sprague straddled the original route of Highway 10 but lost much of its vitality after being bypassed by the construction of Interstate 90 in the 1960s. (WSDA, 008-0807.)

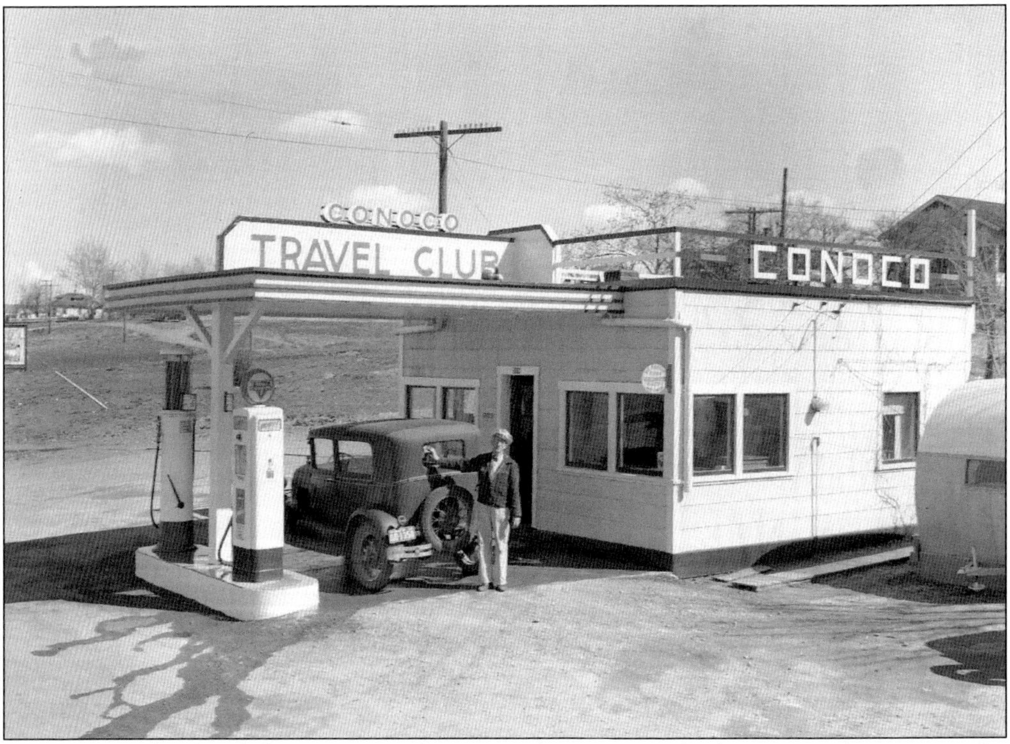

Ritzville was the first full-service, traveler-oriented town since Spokane, 60 miles away in the rearview mirror. In the 1940s, a dozen service stations, restaurants, and tourist courts lined 1st Avenue, the highway's route through town. The Conoco Travel Club, at the north end of 1st Avenue, was operated by Fred Jingling, shown here cleaning the back window of a customer's auto. (WRH, waRitz3041.)

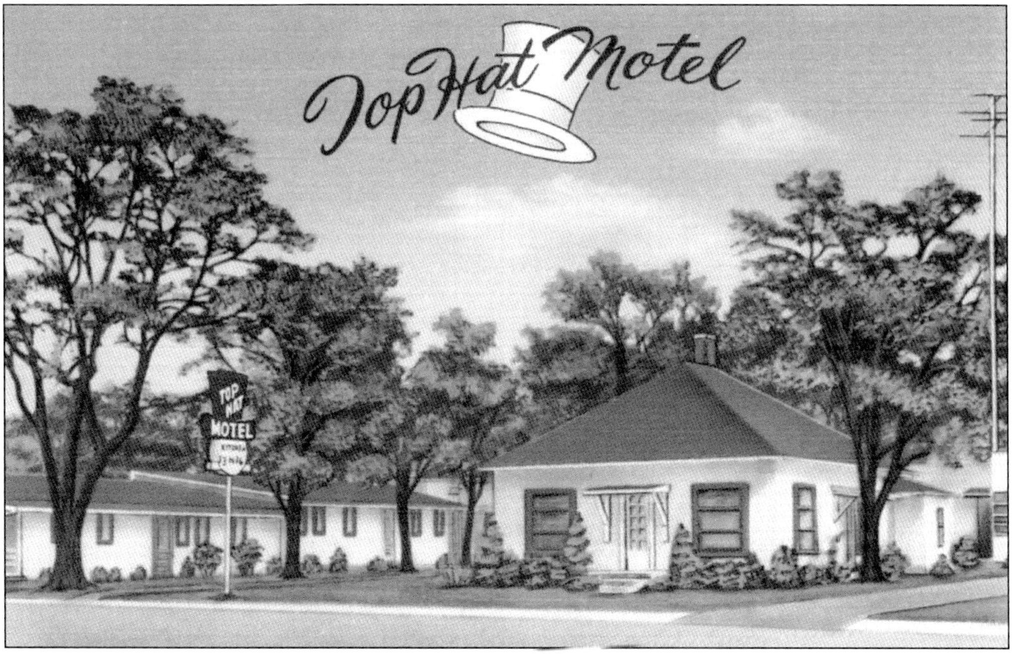

Also at the north end of town was the Top Hat Motel with 11 air-conditioned units, each equipped with kitchen facilities, steam heat, showers, wall-to-wall carpeting, and free television. Room rates were $3 for a single and $6 for a double in 1962 when the motel was owned by Ray and Inez Kilgore. Somewhat modified over the years, the Top Hat is still in business at 210 E 1st Avenue.

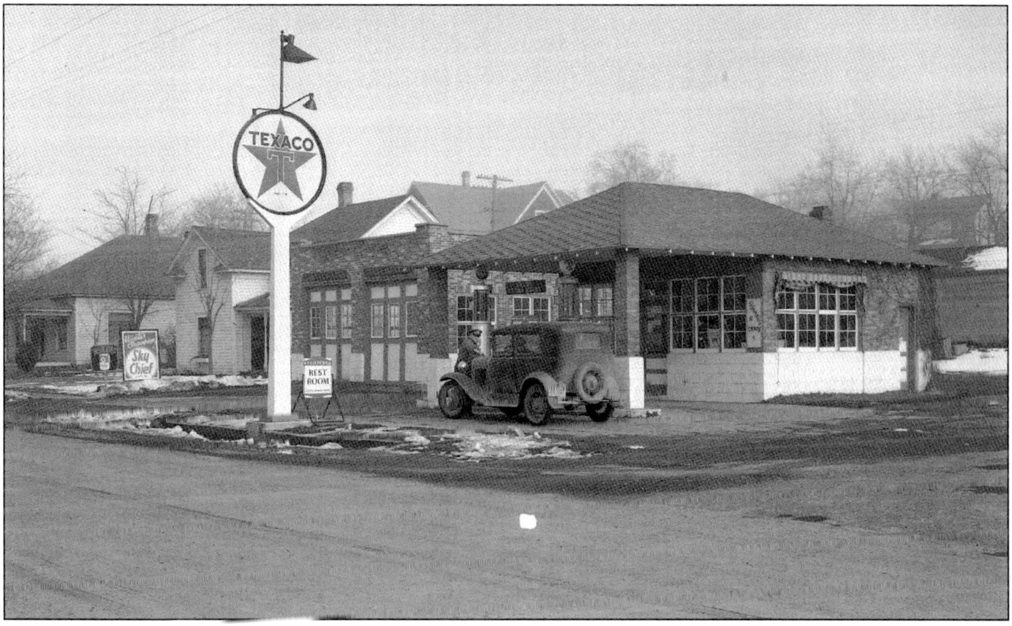

Milner's Texaco Service was an elegant affair: brick built with a decorative band of white concrete blocks, a portico, a three-pump gasoline dispensary, and a two-bay garage which, according to the signs over the doors, offered "Certified Lubrication" and "Tires & Batteries." Even the restrooms were posh. "Registered," says the sign on the sidewalk. Milner's was located on 1st Avenue at Columbia Street. (WRH, waRitz3000.)

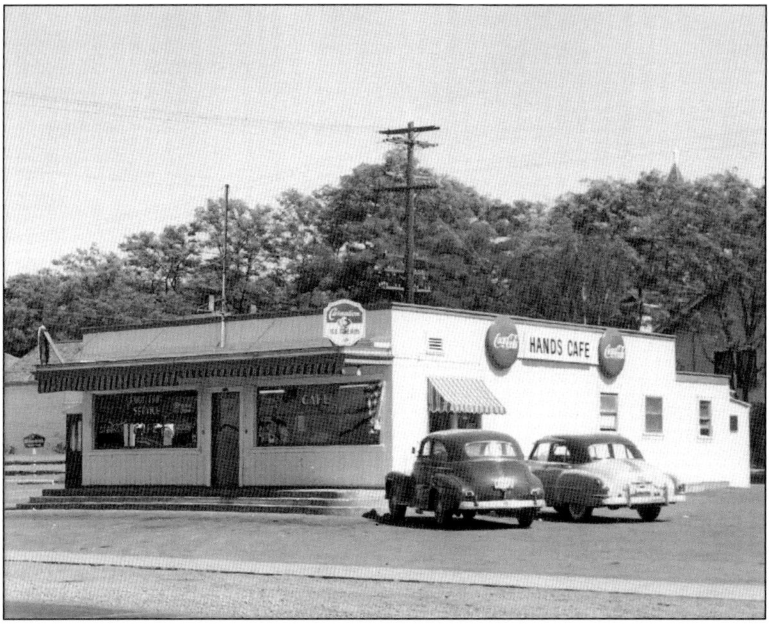

Claude Hand operated Hand's Café from 1949 until 1962 when he sold the business to Bob and Opal O'Brien and moved to Spokane. The café, with "Foods Fit for a King," specialized in steaks and chops; fountain service featured Carnation ice cream. Hand's was also the local Greyhound bus depot. The site, at 110 E 1st Avenue, is now a parking lot. (WSDA.)

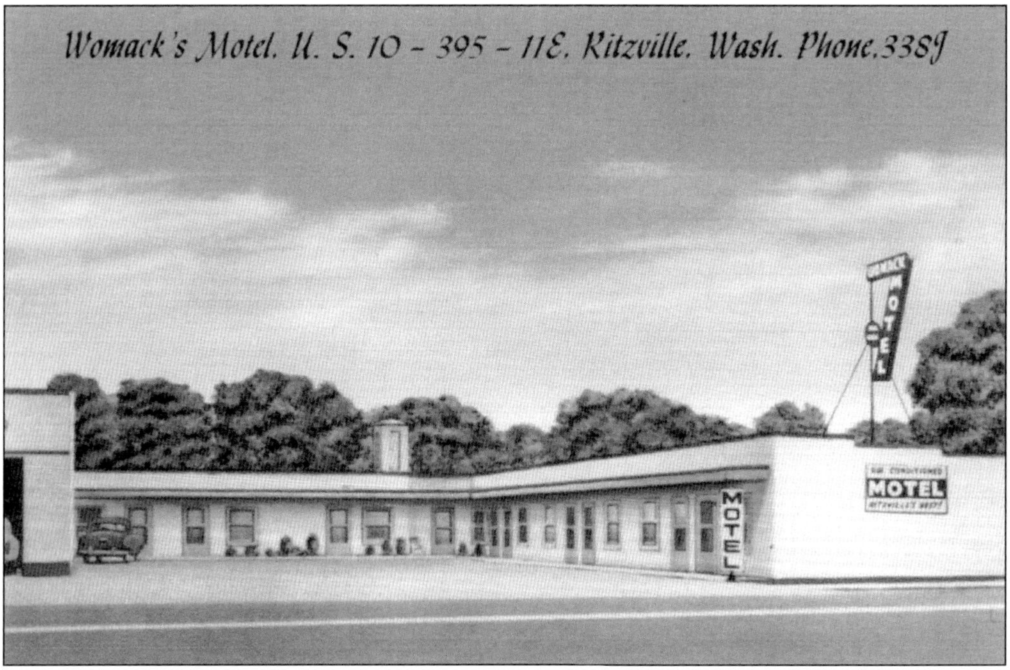

Womack's Motel was across Division Street from Hand's at 105 W 1st Avenue. Its 17 single and double units were arranged in a U shape around an open parking area. Each air-conditioned unit was equipped with Beautyrest beds, showers, and individually controlled central heat. Mr. and Mrs. Charles Womack sold their motel to Mr. and Mrs. Jay Salkeild in 1959. It now does business as the Empire Motel.

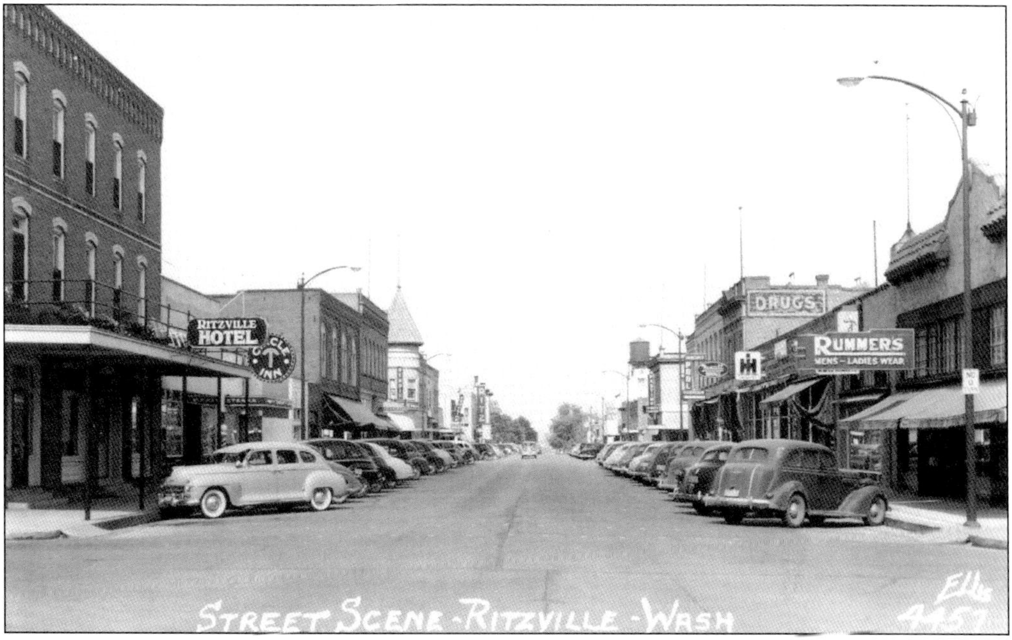

The seat of Adams County, Ritzville, had a 1939 population of 1,748. Its downtown is located two blocks west of Highway 10 and was easily accessible by travelers. Among the many signs to be seen in this view are the Ritzville Hotel, the Circle T Inn, Rummer's Mens and Ladies Wear, Rexall Drugs, the Pastime Sports Center (still in business), the Whisperin Palms, and the Ritz Theatre. (JCO'D.)

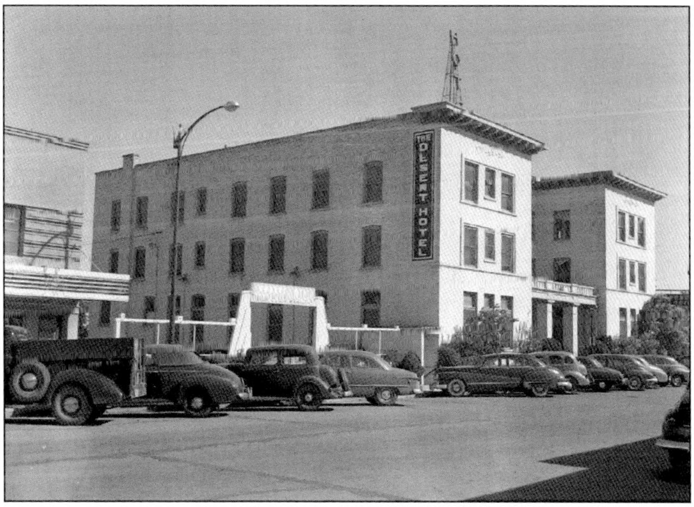

Originally known as the Adams House, the Desert Hotel was acquired by the Dessert family in the 1930s. The Desserts also owned hotels in Spokane and Coeur d'Alene. The hotel was the focal point for many Ritzville social gatherings. It was later known as the Davis Hotel, under ownership of Paul Davis. Located at the north end of Main Avenue, the building was destroyed by fire in the 1980s. (WSDA.)

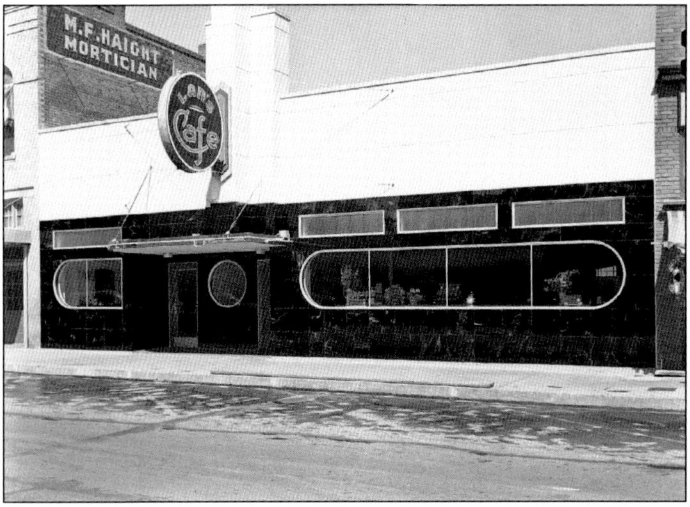

Landreth "Lan" Evans opened a hole-in-the-wall café at 110 W Main Avenue around 1912. By 1939, this modernistic building had taken its place. Lan's was renowned for tasty dishes, generous helpings, courteous service, and modest prices. In 1945, the café was sold to the Fix brothers, Howard and Harold, who operated it as Fix's Café until selling it the following year. It later became the Whisperin Palms. (WSDA.)

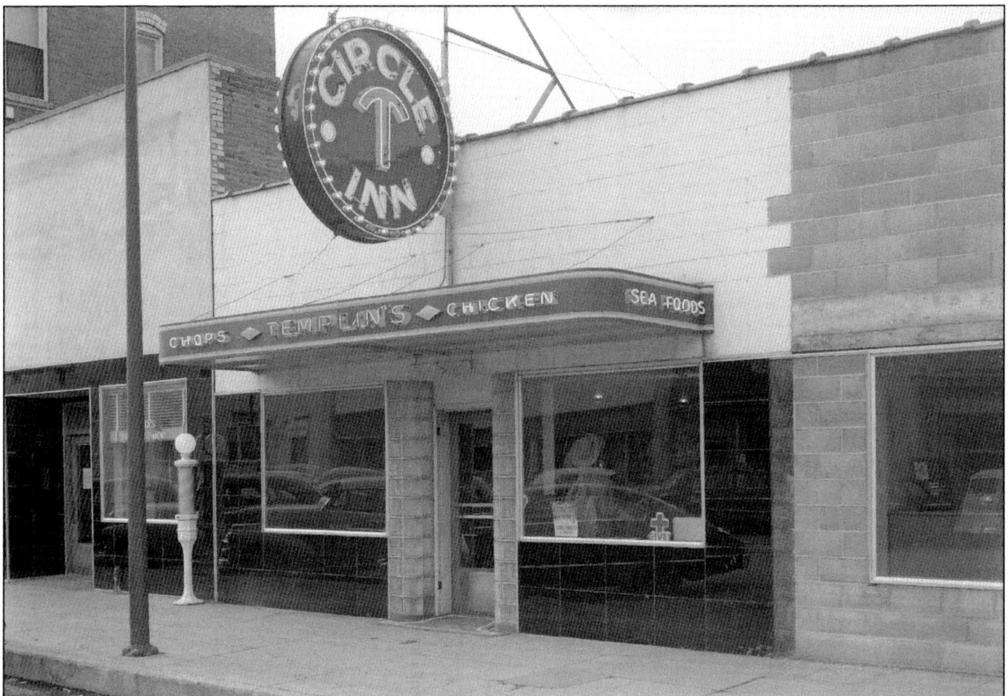

The Circle T Inn was located at 214 W Main Avenue in downtown Ritzville. Open for breakfast, lunch, and dinner, the Circle T ("T" for the owners, the Templin family) specialized in steaks and pan-fried chicken; seafood was also on the menu. A cocktail lounge, the Ranch Room, operated next door. Sold by the Templins in the mid-2000s, the building still houses a restaurant. (WRH, waRitz3011.)

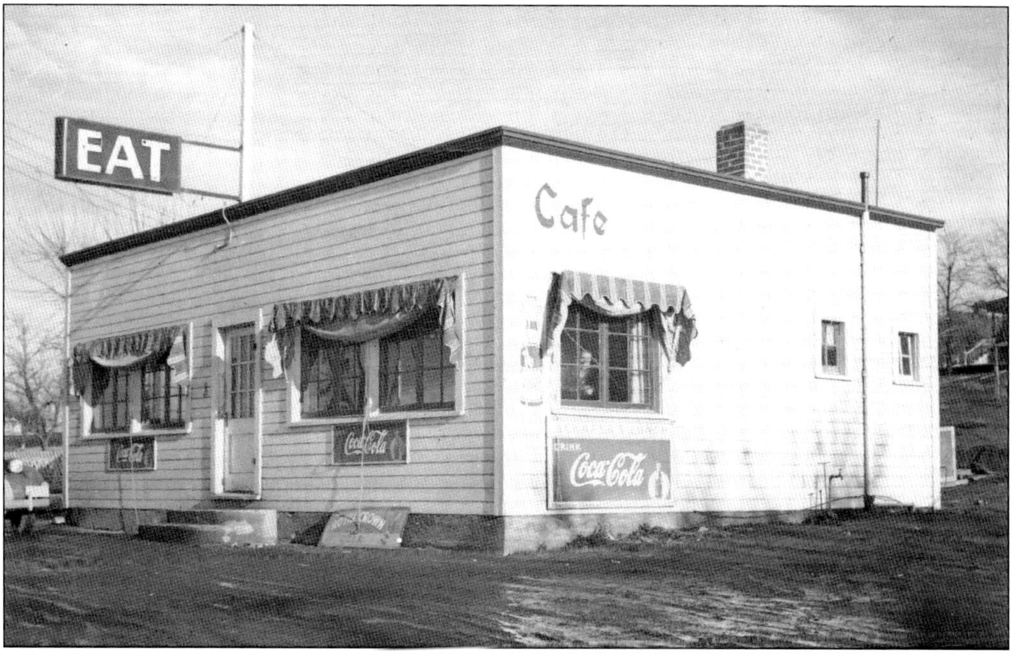

The parking lot at Schafer's Lunch was a muddy mess in this 1939 photograph. When it came to colas, Mr. Schafer was open-minded: signs advertising Coca-Cola, Pepsi-Cola, and Royal Crown Cola adorned his walls. Canvas awnings over the windows—folded up for the winter—add a nice decorative touch to the otherwise plain structure. Note the curious man looking out the window at the photographer. (WSDA.)

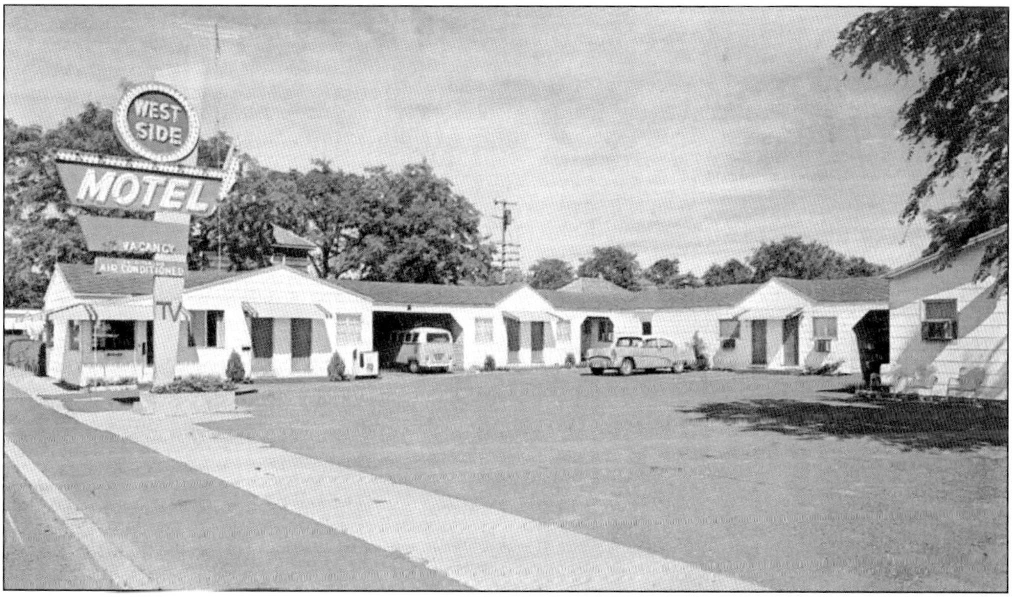

True to its name, the West Side Motel was located on Ritzville's west side at 407 W 1st Avenue. With 10 units, some with kitchenettes, all with Frigidaire air conditioning and television, the West Side was owned by James M. and Rae Phelts in 1960. Though the classic sign is gone, the building still stands. The carports between the units have been enclosed to provide additional housing. (JCO'D.)

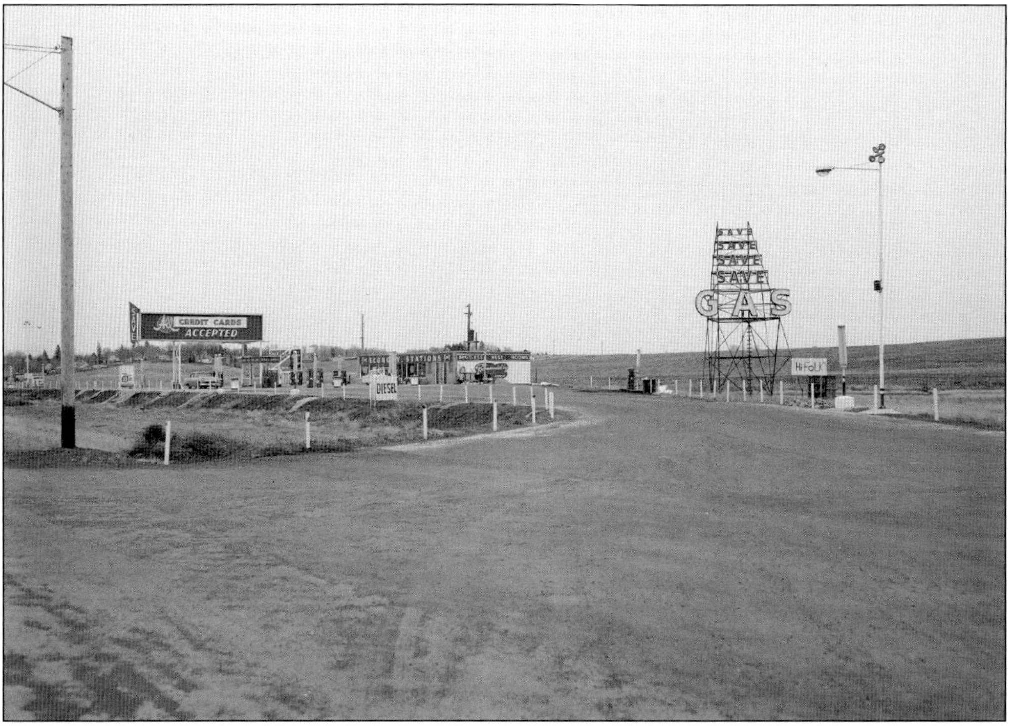

Many residents worried about the impact on downtown businesses—particularly those dependent on travelers—when Ritzville was bypassed by new highway construction in 1959. Some, however, decided to migrate to where the traffic was. The Economy Station, a full-service truck stop, was easily accessible by the new freeway off-ramp south of town. Managed in 1967 by a man named Flyckt, it had been demolished by 2007. (WRH, waRitz3067.)

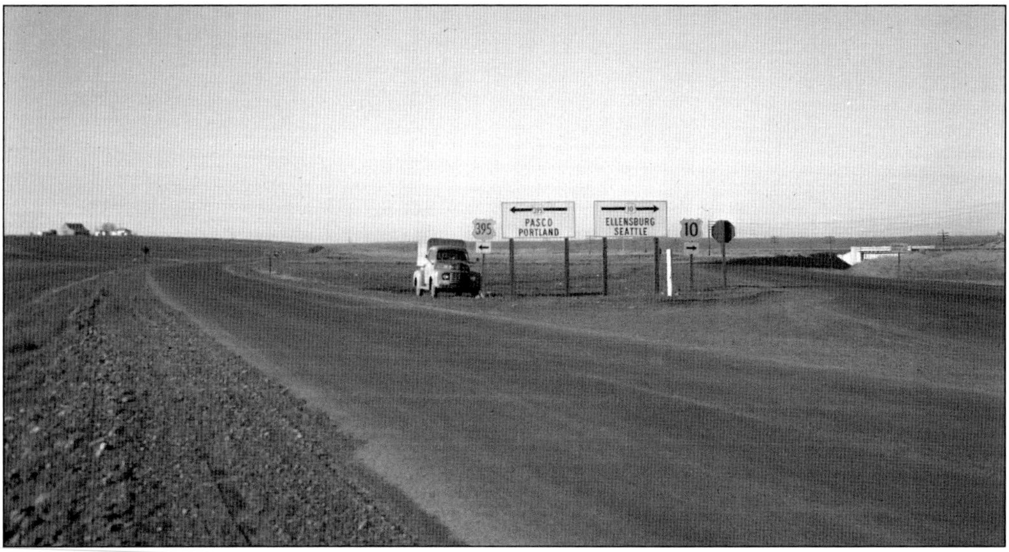

Just outside of Ritzville was a major highway junction: Highway 395, the old Central Washington Highway, headed southwest to Pasco and on to Portland, Oregon, while Highway 10 curved west towards Moses Lake and Seattle. For Highway 10 travelers, beyond this point were rolling hills of wheat and virtually no services until Moses Lake, 40 miles away. (WSDA.)

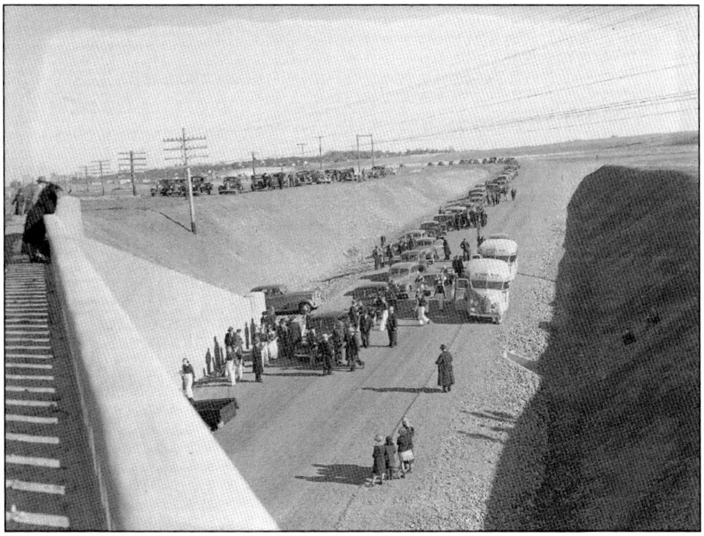

Dedicating an underpass: prosaic to some but a major event for the people of Ritzville, and a large crowd turned out on November 8, 1939, for the occasion. Ritzville is bisected by the railroad—downtown is west of the tracks, the highway is east; railroad traffic through town often cut the two sides off, and the new underpass made it possible to travel around stopped trains. (WRH, waRitz3091.)

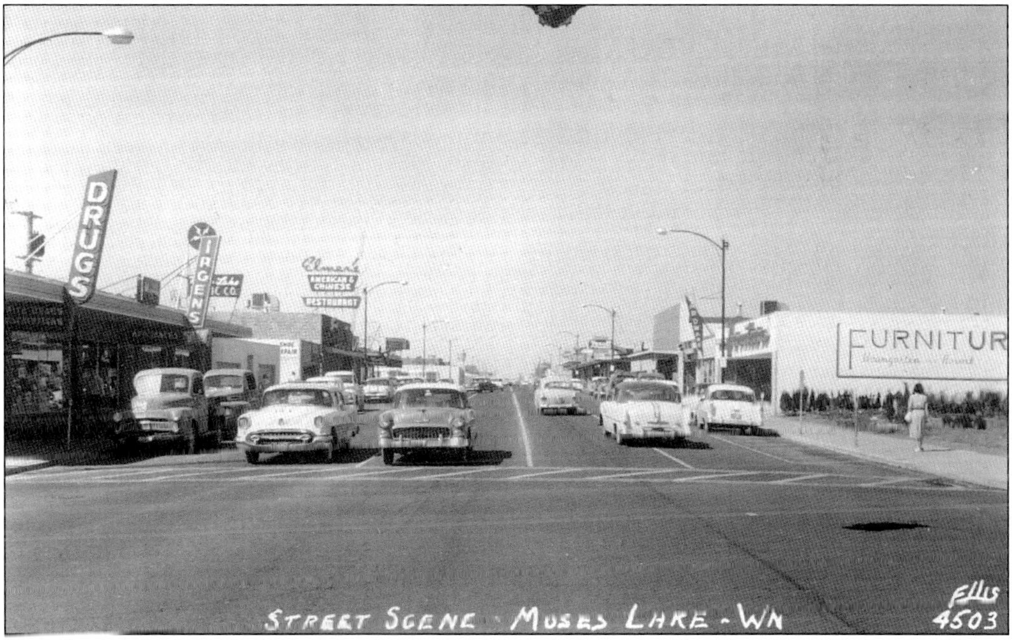

In early years, Moses Lake (originally named Neppel) was an isolated spot in Washington's vast Big Bend country. The city was not incorporated until 1938. When State Route 18 was adopted as the new alignment for Highway 10 between Ritzville and Vantage in the late 1940s, Moses Lake found itself on the state's major east-west highway and Broadway Avenue became lined with motorist-oriented businesses and services.

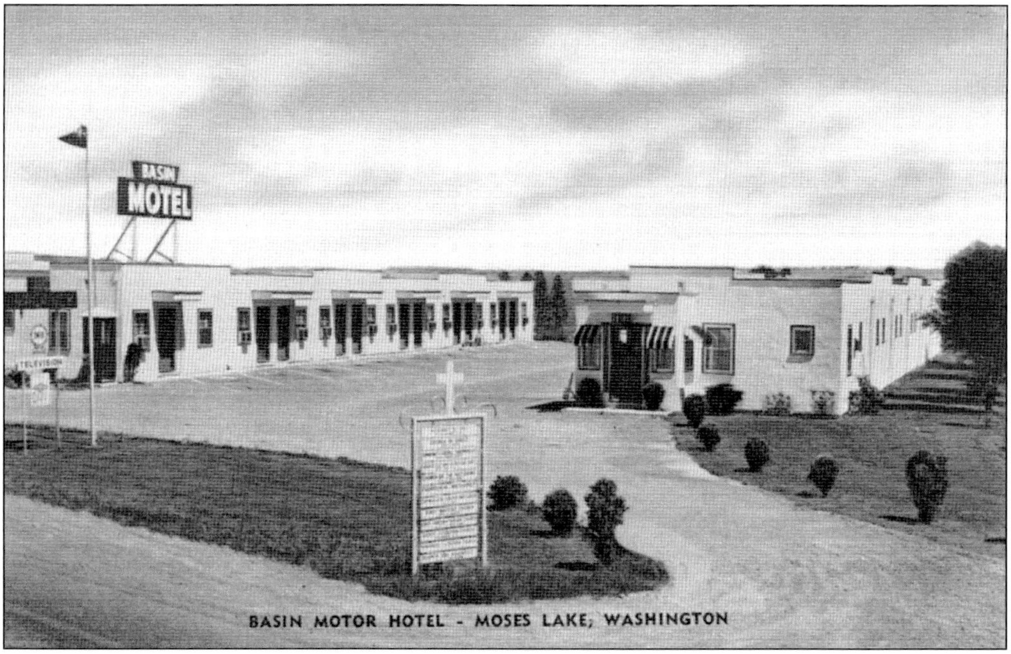

Bob Trask Sr. moved to Moses Lake in 1946 when the city's population was only 700 (it is now over 20,000) and opened the Basin Motor Hotel along with an auto dealership, a construction operation, and an insurance agency. The 24-unit motel was located east of town on Highway 10. Heavily damaged by fire in June 1950, the motel was rebuilt and sold to Elbert Goodnight in 1955.

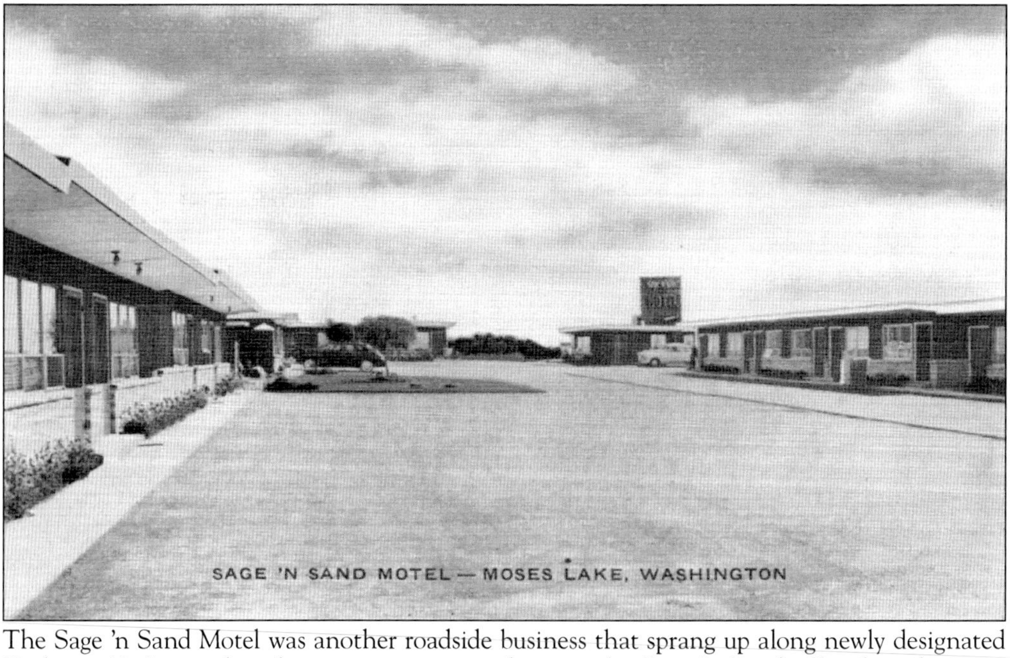

The Sage 'n Sand Motel was another roadside business that sprang up along newly designated Highway 10. Located at the east entrance to the city, the Sage 'n Sand offered modern air-conditioned sleeping rooms and apartments with kitchenettes, tile showers, and radios. The motel, much expanded and updated, is still in business at 1011 S Pioneer Way in Moses Lake.

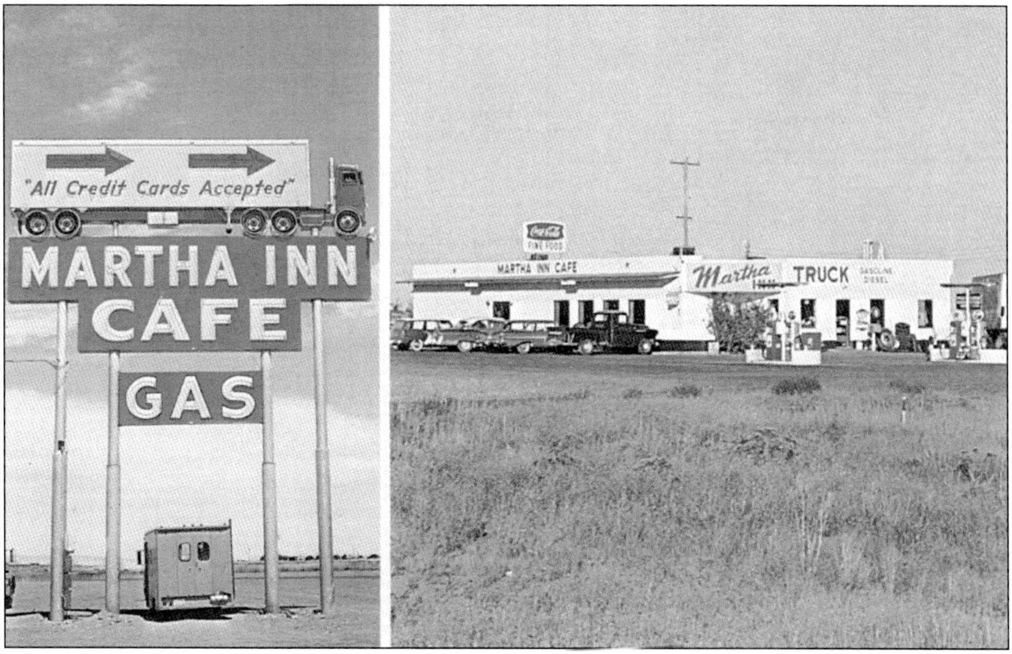

The Martha Inn was located, fittingly, in the town of George, Washington. George was a planned community developed in the mid-1950s adjacent to Highway 10 in a relatively unpopulated area of the vast Columbia Basin Plateau. Located midway between Seattle and Spokane, the Martha Inn attracted truckers with year-round 24-hour gas and repair service and was known for its fine food. (JCO'D.)

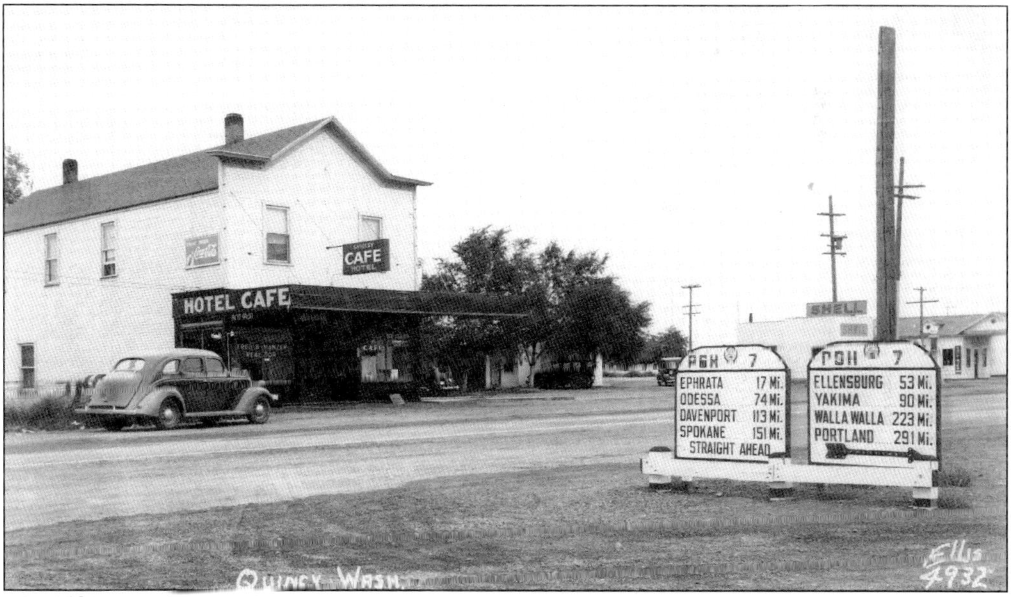

The 1930s travelers sought to avoid the troublesome Blewett Pass route by opting for Primary State Road 7, a highway linking Davenport, Quincy, and Ellensburg via a crossing of the Columbia River at Vantage. Quincy became a travel hub with service stations, cafés, and the Shirley Hotel. Seen here in 1946, the hotel had been recently repainted and reopened by owners Mr. and Mrs. J.W. Adams. (WSHS, 2013.154.1.)

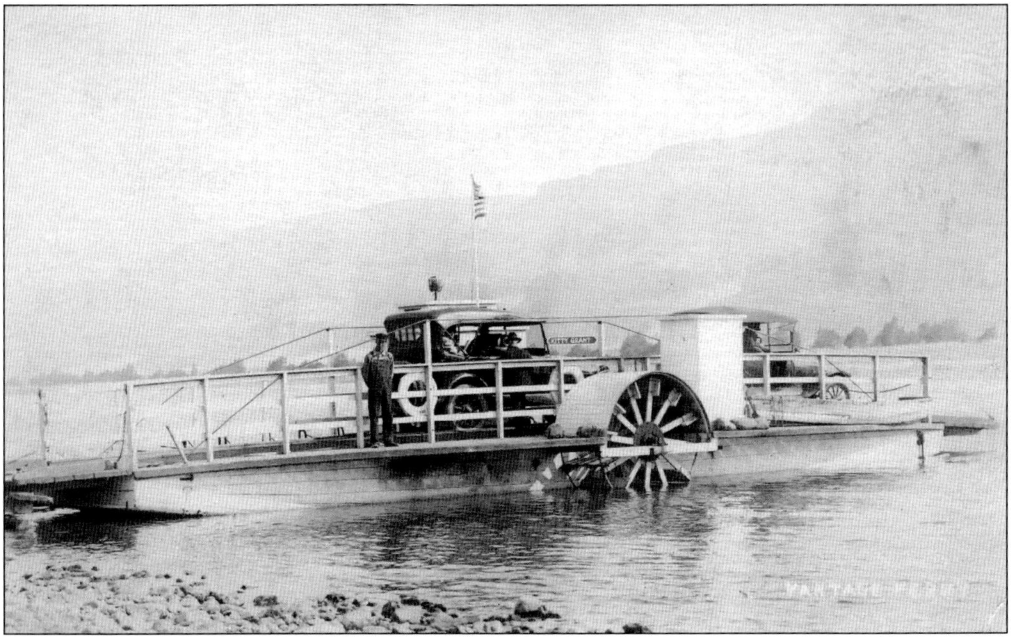

The Vantage ferry crossing of the Columbia River dates to 1914, shortly before Primary State Road 7 was decreed into existence. However, it was not until the Vantage Bridge was opened in 1927 that the Quincy-Vantage-Ellensburg route became a feasible alternative to Blewett Pass. This ferry, the *Kitty Grant*, transported vehicles across the river from 1917 until construction of the bridge. (KCHM, 82-23-3.)

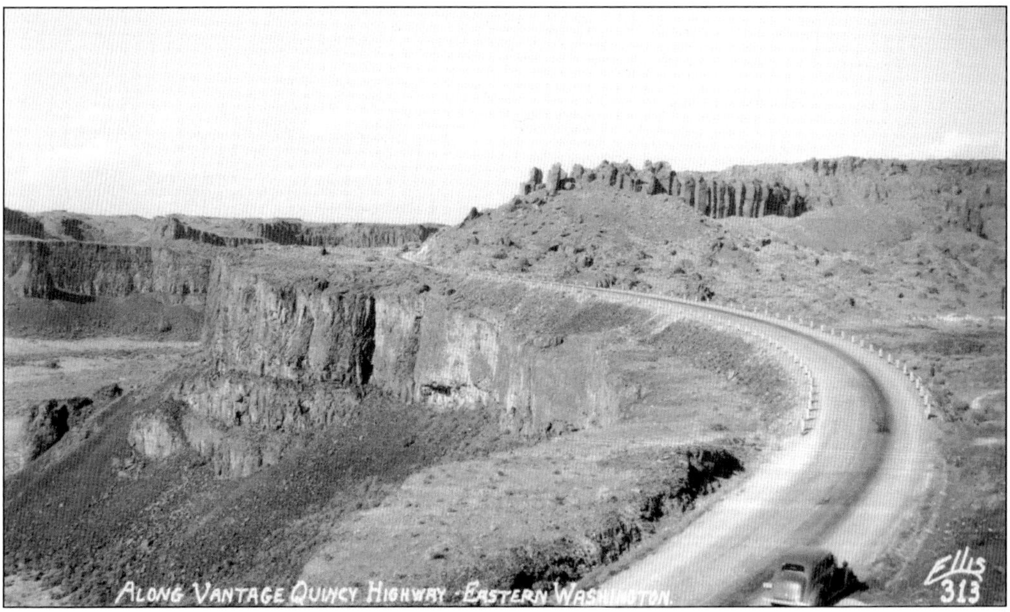

For years, Primary State Road 7 was nothing more than a primitive road between Quincy and Ellensburg—a highway in name only. The grade between Ellensburg and the river crossing at Vantage was manageable, but steep bluffs east of the Columbia seriously impeded travel. Construction projects gradually improved the route, particularly the climb up from the Columbia. East of the river, the highway was known as the Vantage-Quincy Highway.

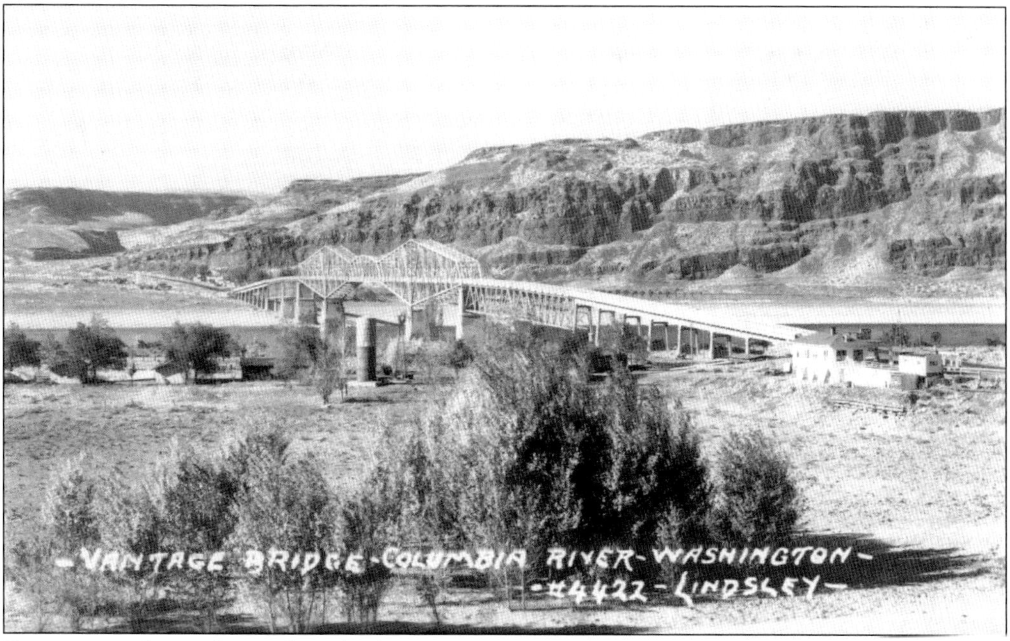

Recognizing the need to replace the ferry at Vantage, in 1927, a two-lane bridge was opened to serve ever-increasing traffic, becoming part of the 1940s realignment of Highway 10 through Ritzville to Ellensburg. Construction of Wanapum Dam in 1962 required the bridge to be replaced by a new crossing a mile south; the old bridge was recycled to cross the Snake River at Lyons Ferry.

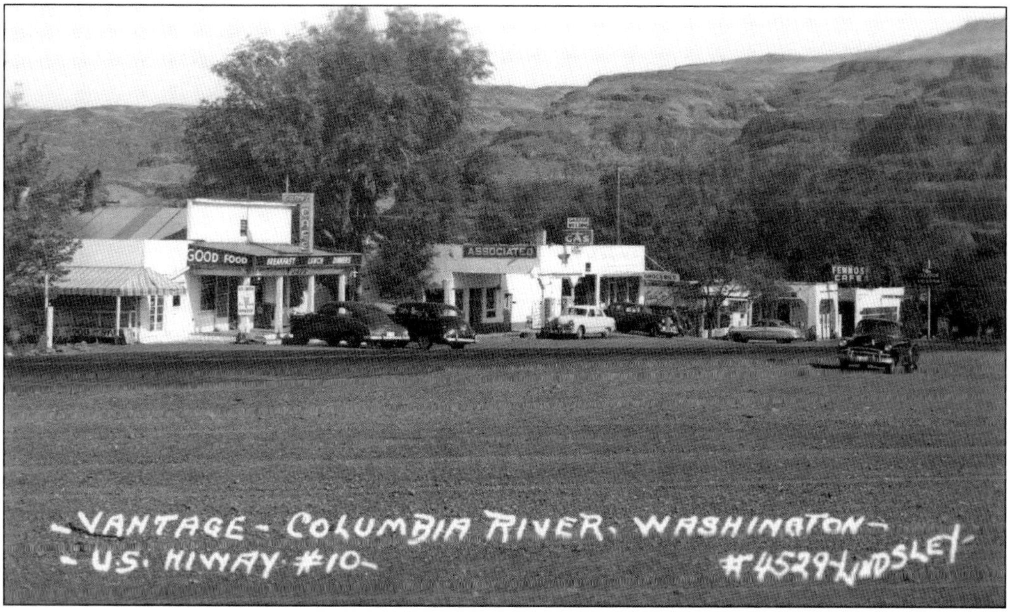

Roadside businesses popped up at both ends of the Vantage Bridge—a café and Texaco station on the east bank, and Stockdale's Tourist Camp, two cafés (Gray's and Fenno's) and an Associated gas station on the west bank in the actual settlement of Vantage. The sites of all these buildings are now under a hundred or more feet of water backed up behind Wanapum Dam. (WSHS, 2011.80.3.)

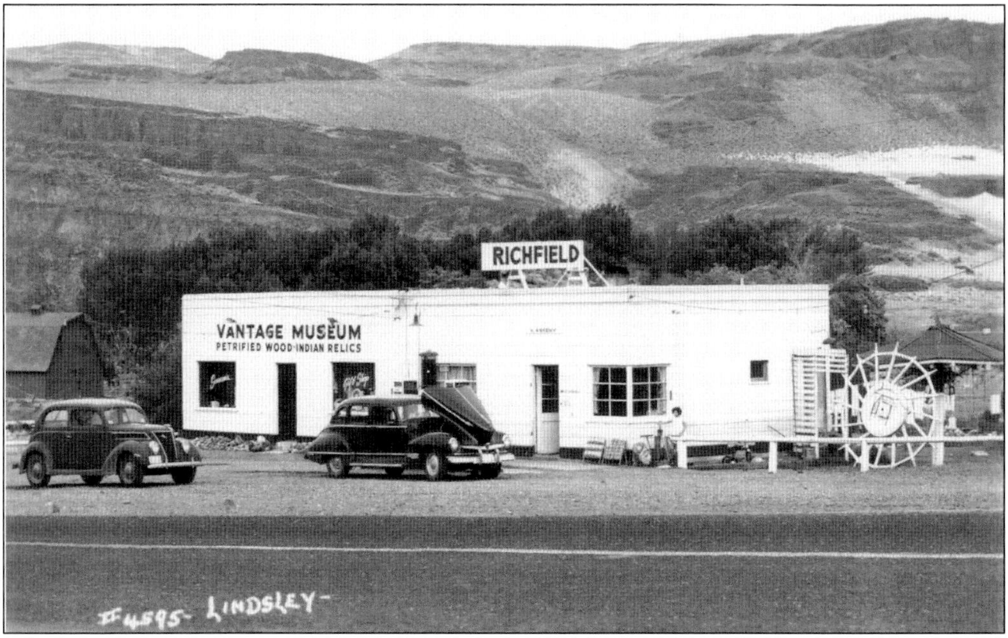

Vantage was an archaeologically and geologically rich area. The riverbanks yielded artifacts of the original inhabitants, pictographs lined the canyon walls, and rock hounds found petrified wood in the surrounding hills. In the late 1940s, Harry Ankeny opened this private museum at his service station to display many of the relics he had found. Backwaters of Wanapum Dam now cover the site, though many pictographs were salvaged.

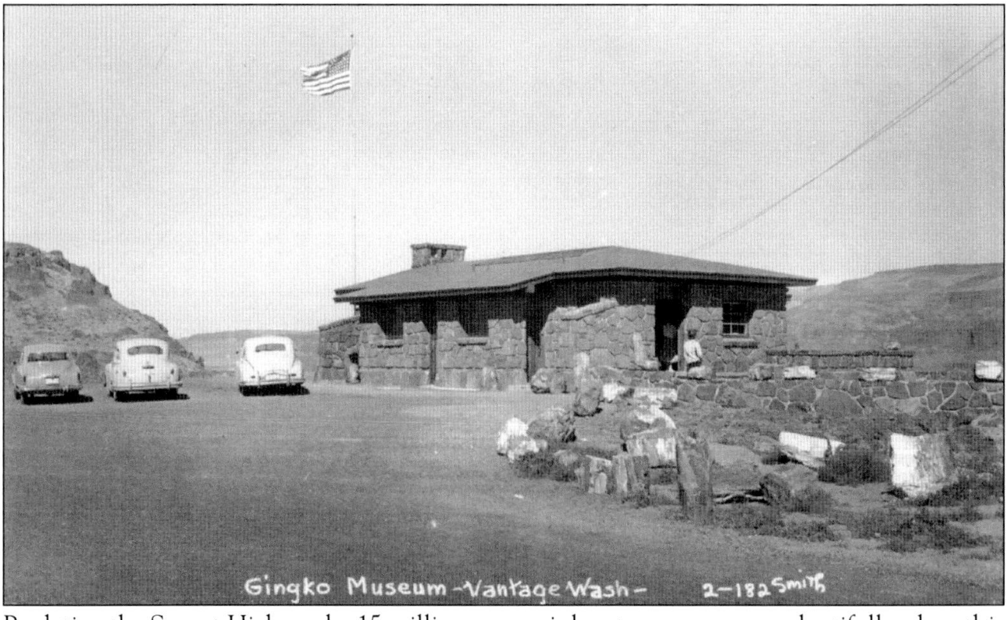

Predating the Sunset Highway by 15 million years, ginkgo trees once grew plentifully when this now-arid region was a lush wetland. After fossil trees were discovered by highway workers in the 1920s, a state park was created to preserve and protect them. A visitor center and museum was constructed along with a 1.5-mile walking trail. Though Interstate 90 rerouted most traffic to the south, the park is still often visited.

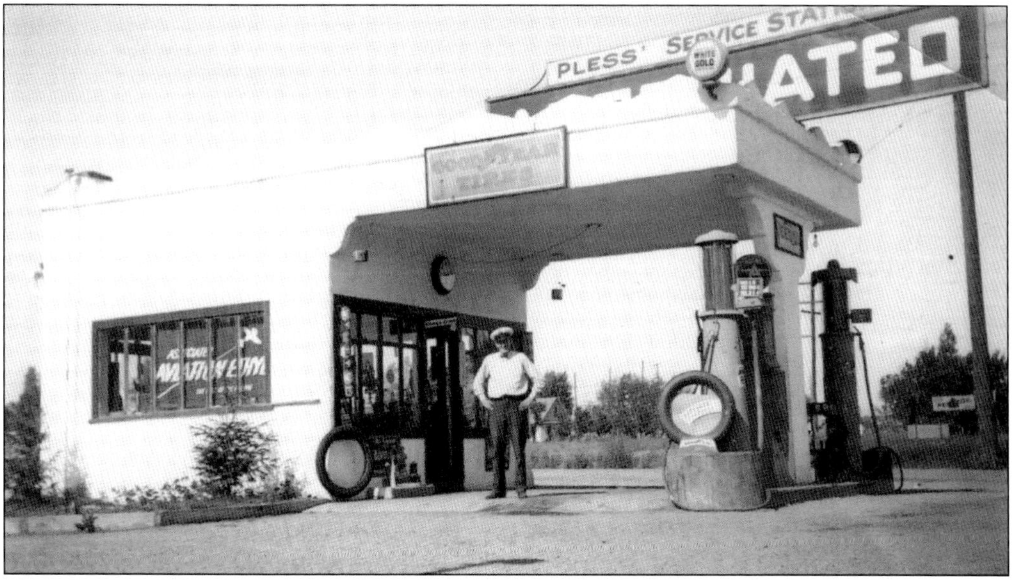

Pless' Service Station was located at 8th and Chestnut Streets in east Ellensburg. By 1939, it was known as Bun's Service Station ("Bun" was owner Bernard DeShaw), though an associated tourist court continued to be called Pless' Auto Camp for a while after the station changed hands. Product allegiance changed too—Pless sold Associated gas and Pennzoil; Bun favored Signal gas and Quaker State motor oil. (KCHM, 91-29-1.)

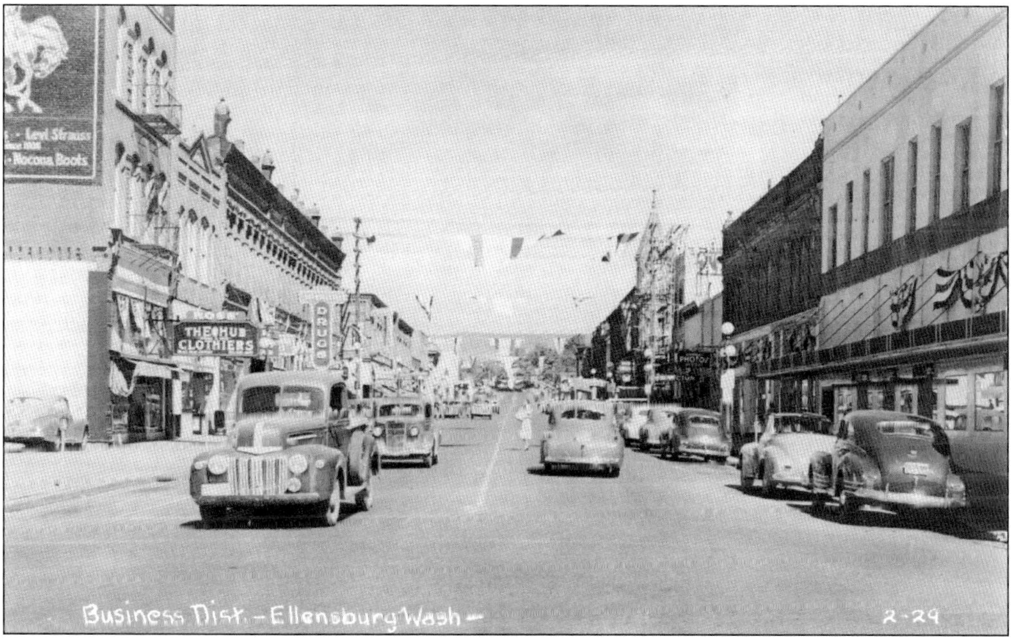

Highway 10 passed about a quarter of a mile north of downtown Ellensburg, seat of Kittitas County. At the intersection of 8th Avenue (now University Way) and Main Street, motorists could turn towards the business district with its many stores, hotels, and restaurants. Main Street (Highway 97) carried southbound traffic out of the city into the scenic Yakima River Canyon, then to Yakima and on towards Oregon.

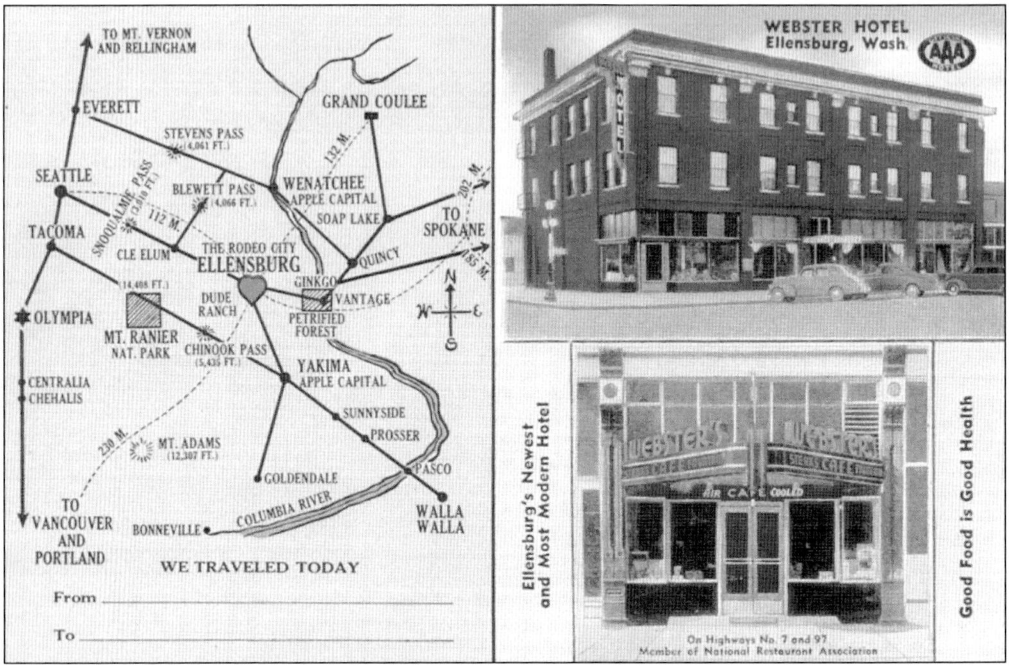

The Webster Hotel was at the heart of everything, according to the map on its advertising card. Located at Third and Pearl Streets in downtown Ellensburg, the hotel was built in 1909 as the Majestic. In 1937, it became the Webster when purchased by William Webster. The hotel offered modern rooms at reasonable rates, and Webster's Café claimed the "finest in foods, properly served." Fire destroyed the building in 1980.

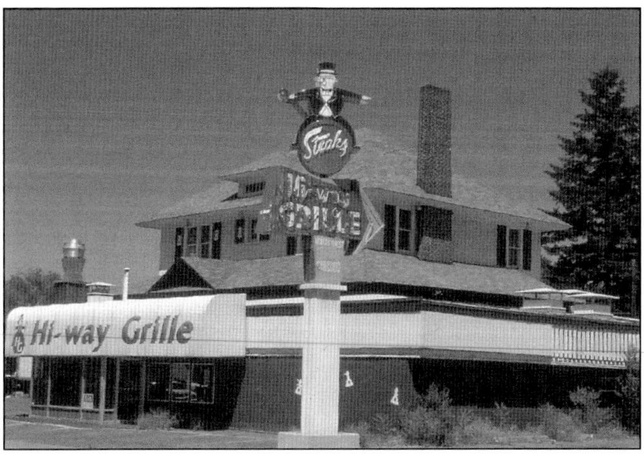

Vivian Sorenson-Martin opened the Hi-Way Grille in 1940 on E 8th Avenue, moving a few blocks west to the junction of Highways 10 and 97 in 1957. A long-time institution among townspeople, travelers, and college students, the restaurant was famous for its specialty, "the World's Finest Eatin' Chicken." Steaks were also a featured item, with cocktails available in the Galaxy Room. A fast-food franchise now occupies the site.

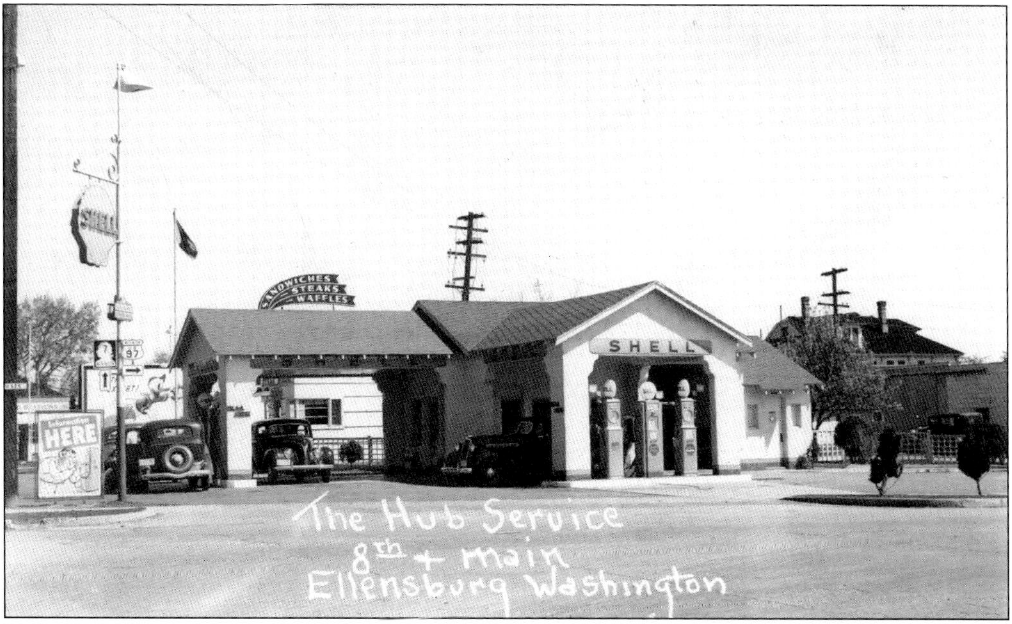

Hubert "Hub" Holmes opened this service station prior to 1937 at the southeast corner of 8th Avenue and Main Street, the busy intersection of the Vantage Highway and the route south to Yakima. With two three-pump bays, The Hub could easily handle traffic rolling in to fill up on Shell gas. A sign for The Steak House looms above the left bay; another sign depicts Donald Duck. (EllensburgPL, BBS383.)

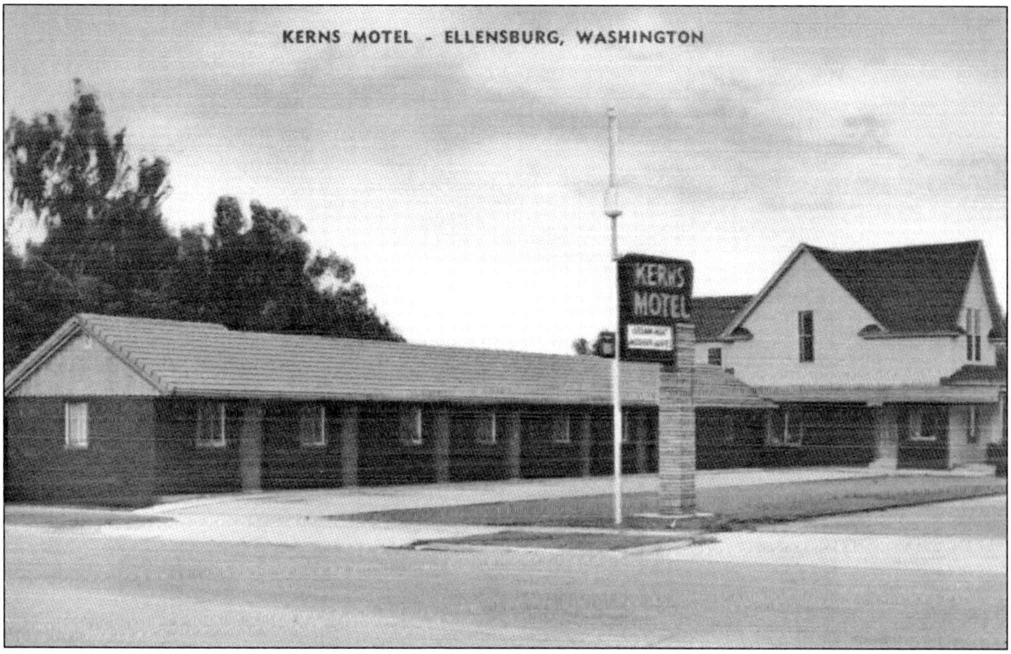

Also at the Highway 10-97 junction was Joe and Dorothy Kerns' motel at 709 N Main Street. Its tile-roofed brick units offered cleanliness, comfort, and hospitality along with steam heat, free radios, and ice. Commercial rates were available for frequent travelers. Kerns Motel has long since disappeared; the location is a parking lot today.

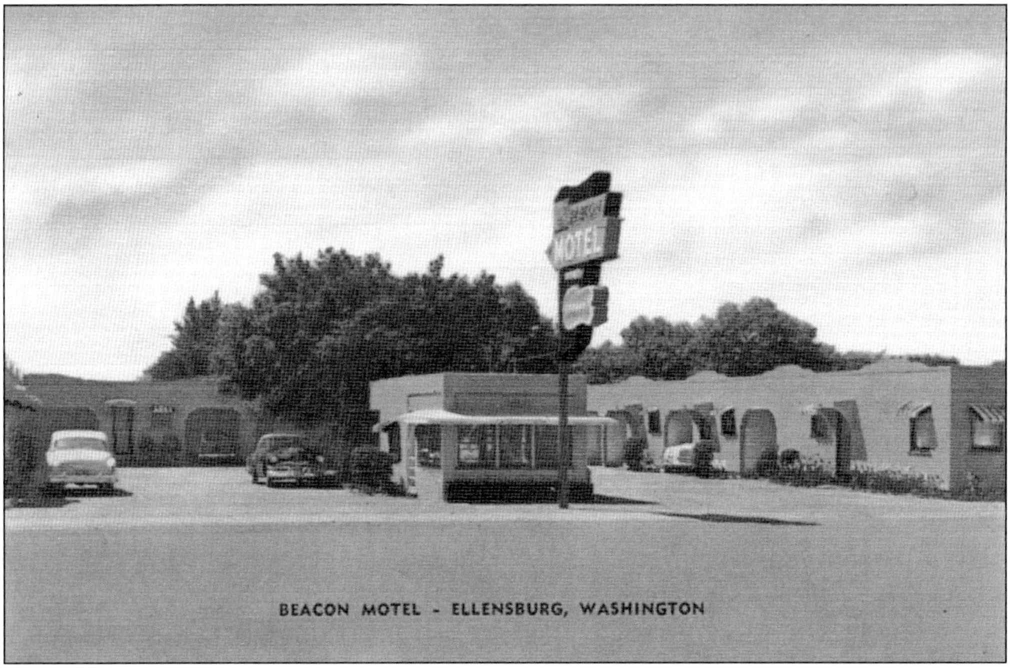

Resplendent in aquamarine paint, the Beacon Motel was at 404 West 8th in Ellensburg, just west of where Highway 97 departed Highway 10. The motel had 14 single and double units with steam heat, showers, and kitchenettes. Owned by Mr. & Mrs. O.B. Lamar, it was later run by Mr. and Mrs. L.W. May and, still later, by Hugh and Mary Boucher. The building still exists as the Spur Apartments.

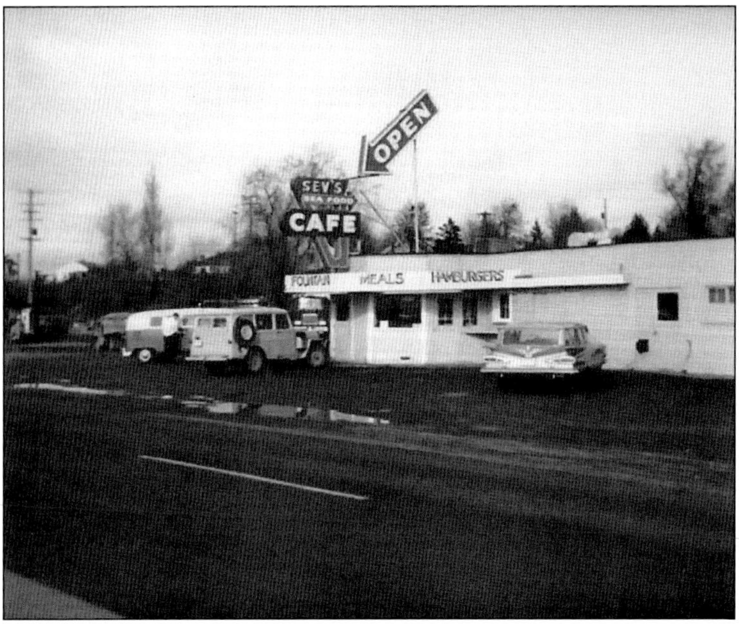

George Bernard "Bernie" Severin opened Sev's Café at 607 Cascade in 1947. He and wife Frances managed the restaurant until 1967. A typical roadside eatery with booth and fountain service, Sev's menu choices ranged from short-order selections—such as hamburgers—to full dinners. Sev's was on the west edge of town; past this point there were not many dining options until Cle Elum, 25 miles away. (KCHM, 2013-181-001.)

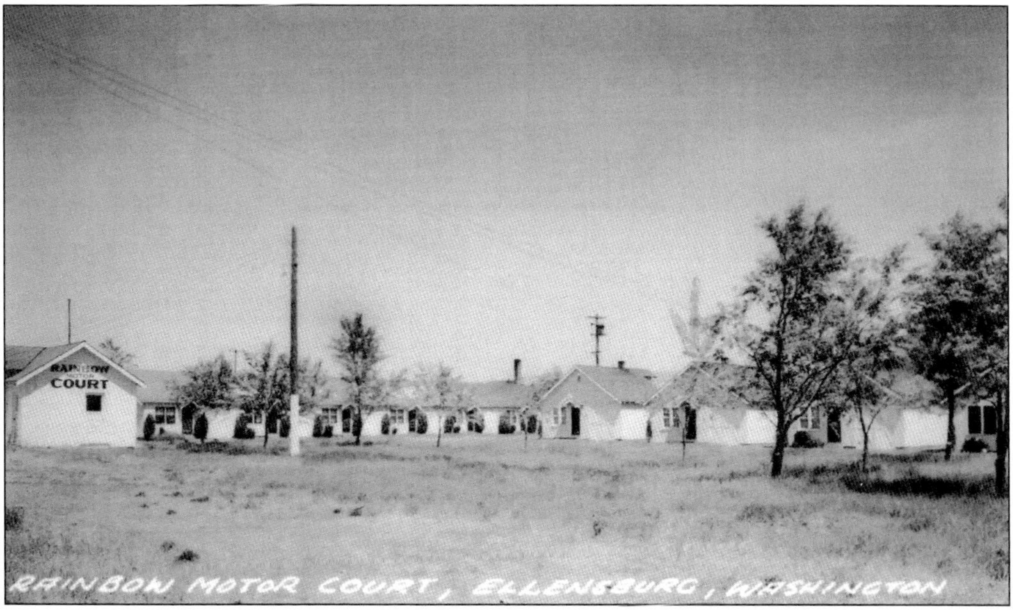

The Rainbow Motor Court opened on the outskirts of Ellensburg around 1940. Consisting of four large detached cabins and a row of 10 single units with an on-site gas station, the Rainbow was operated by Victor Munz in 1942. By 1954, Carmi and Margaret Hanchett were listed as managers. Somewhat altered, the Rainbow is still in business at 1025 W University Way. (EllensburgPL, BBS385.)

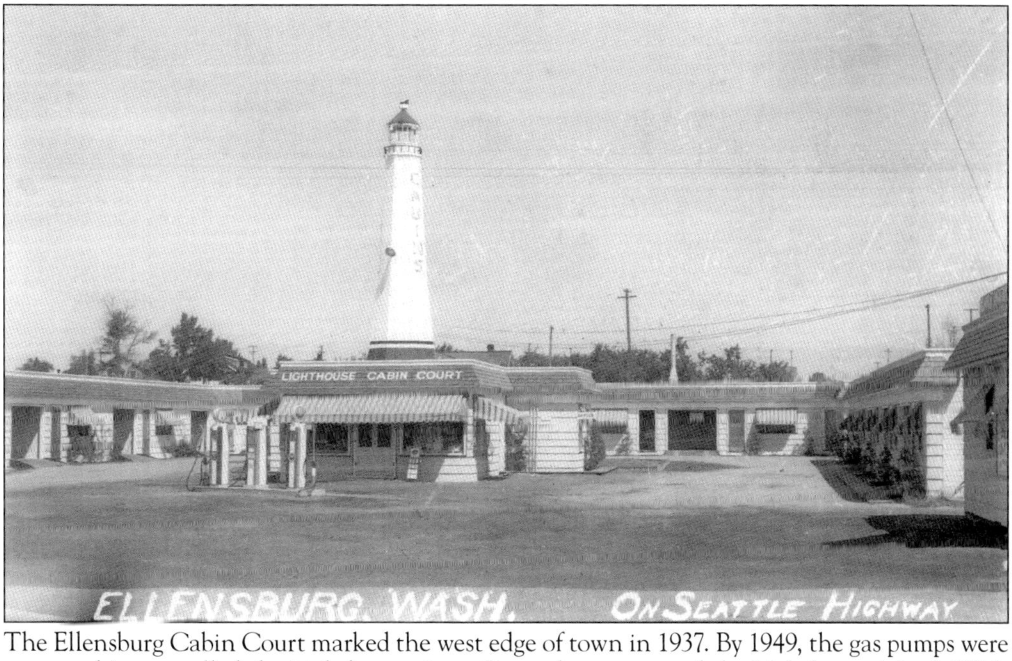

The Ellensburg Cabin Court marked the west edge of town in 1937. By 1949, the gas pumps were gone and it was called the Lighthouse Auto Court, later renamed the Lighthouse Motel. With 16 units, tubs and showers, kitchenettes, and a coffee shop next door, the motel was owned by Hugh and Mary Boucher in 1953. The Bouchers also operated the Beacon Motel up the street. (EllensburgPL, BLD040.)

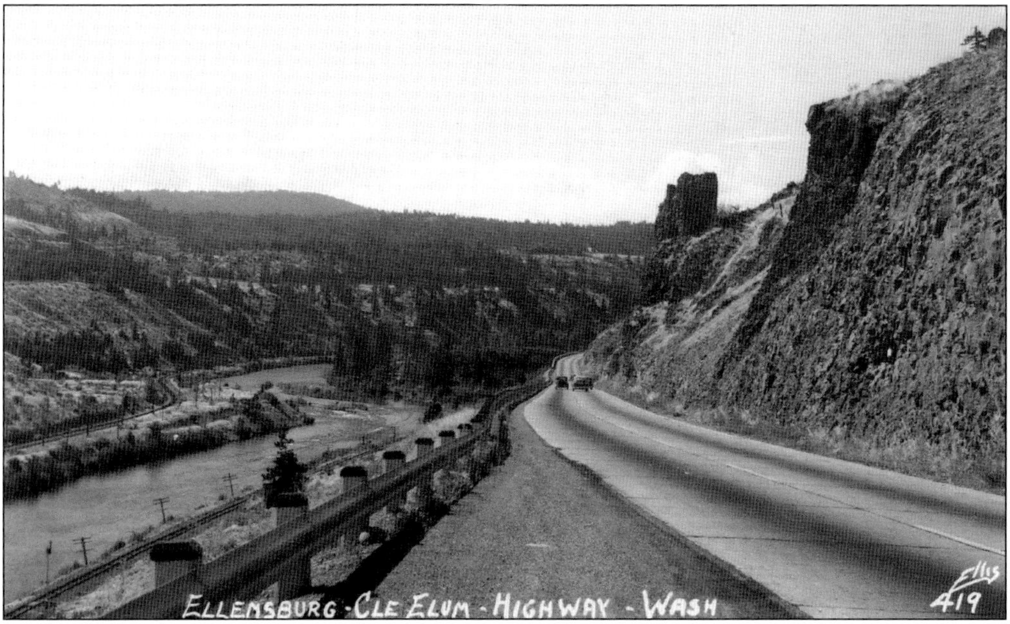

Highway 10 headed northwest up the Yakima River valley after leaving Ellensburg. This relatively recent route was known as the Ellensburg–Cle Elum Highway and was part of US Highway 97 before the Highway 10 realignment in the early 1940s. West of the turnoff to the small town of Thorp the highway enters a scenic canyon, at times rising several hundred feet above the river through rugged rock formations. (EllensburgPL, TRN154.)

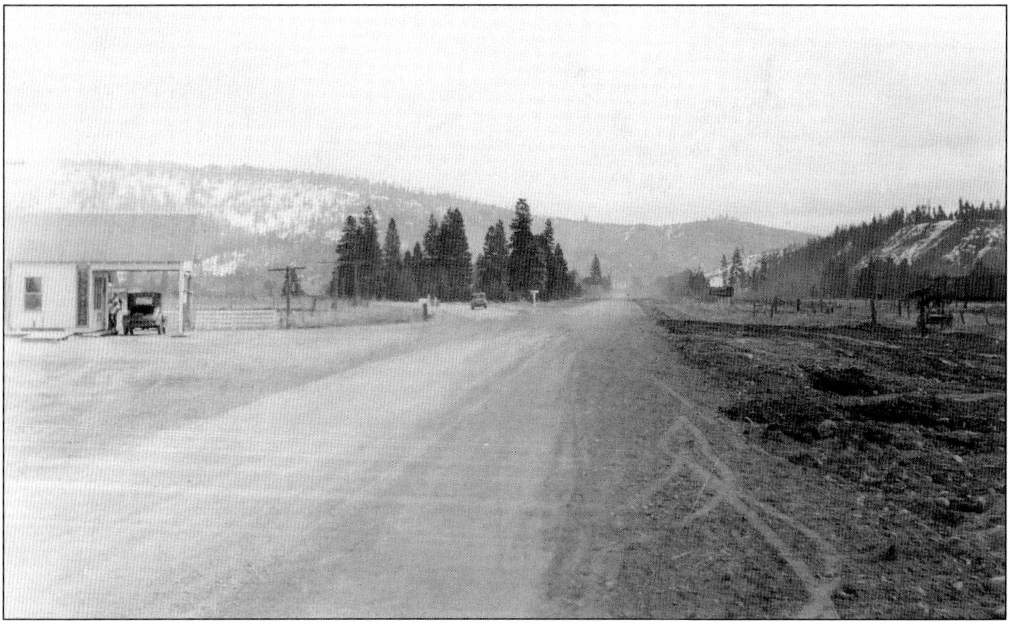

The Sunset Highway's original route reunited with Highway 10 four miles east of Cle Elum at Teanaway Junction. The old road merges from the north (left) while the new alignment out of Ellensburg heads directly towards the camera. At the junction, Platt's Service Station served up Firestone tires and gasoline from two old-style gravity pumps. Tourist cabins and a dining room for light lunches were also available. (WSHS, 2011.0.182.)

Four

Cle Elum to North Bend

With few variations, the Sunset Highway west from Cle Elum has historically followed a single path all the way to North Bend. That route has been dictated by the topography of rivers, lakes, and mountains. Snoqualmie Pass, the most practical passage and lowest point through the mountains, is the determining factor: in use since ancient times, the route across the pass became known in the 1850s to newly arrived settlers to the Puget Sound region. By 1875, a workable road was in place, though needing constant maintenance and closed each winter by deep snow.

Lake Keechelus lies immediately east of Snoqualmie Pass. Dammed today with its water surface about 70 feet above natural level, in early days the lake dictated travel, filling as it did the valley on the eastern approach. Railroads had acquired right-of-way along the west edge of the lake, forcing the nascent wagon road (and, later, the highway) to squeeze along the forested, steep eastern bank.

To the west of Snoqualmie Pass the elevation drops rapidly—1,600 feet elevation loss in six miles west of the summit compared to 500 feet an equivalent distance east. A series of switchbacks accomplishing the drop, the highway emerged into the upper valley of the South Fork Snoqualmie River and followed down the narrow valley to North Bend.

This rugged country was not conducive to roadside businesses. Several hotels and resorts edged Lake Keechelus and there was an inn and garage at Snoqualmie Summit, but Easton provided the last full services until North Bend, 40 miles away. Those could be 40 long, difficult miles for a motorist in trouble or caught by a winter storm.

Modifications to the original route have been few. Highway realignment left Easton off the main route, and a segment of old highway remains at the Snoqualmie Pass Summit. The early switchbacks survive in pristine shape and are still drivable, and a two mile length of original road can be found at Olallie State Park. Elsewhere, the Sunset Highway and its successor, Highway 10, have been overlain by Interstate 90.

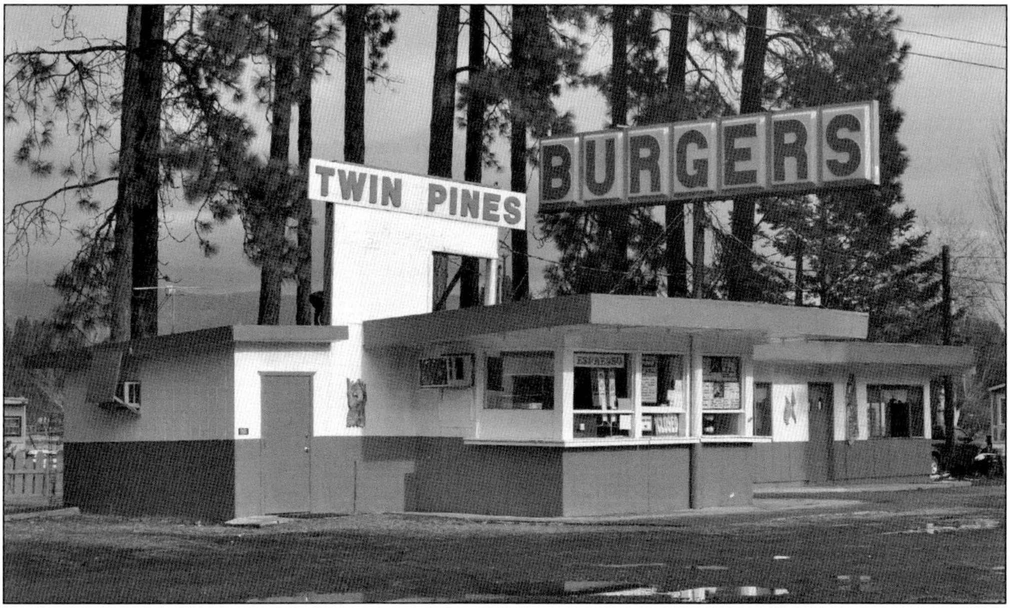

The Twin Pines Drive-In, a highway survivor east of Cle Elum, is as authentic an example of roadside Americana as can be found anywhere. In business for over 50 years, it is road food at its best—burgers, fries, and real milkshakes—with indoor seating and take-away service for enjoying the clean air and blue skies at the picnic tables out back. It is currently only open summers.

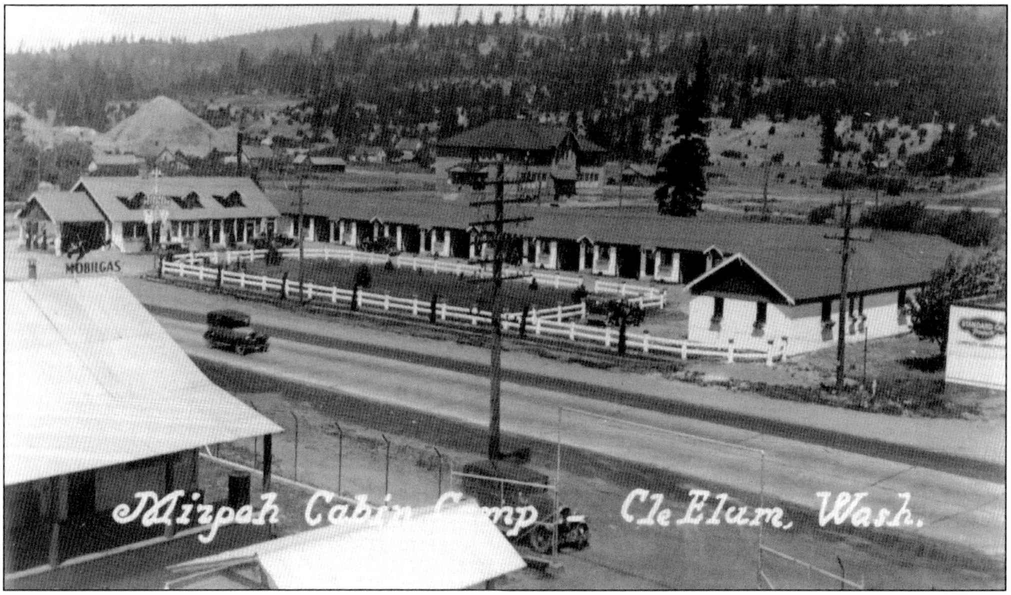

The Mizpah Cabin Camp opened in 1932 in east Cle Elum. Billed as the "latest word in tourist camp construction," the camp consisted of two large structures. One housed an office, store, service station, and manager's quarters for the camp. The other contained the sleeping units, each furnished with a double bed, floor rugs, and kitchenette. A green lawn accented the camp's pleasant and comfortable appearance. (EllensburgPL, BLD061.)

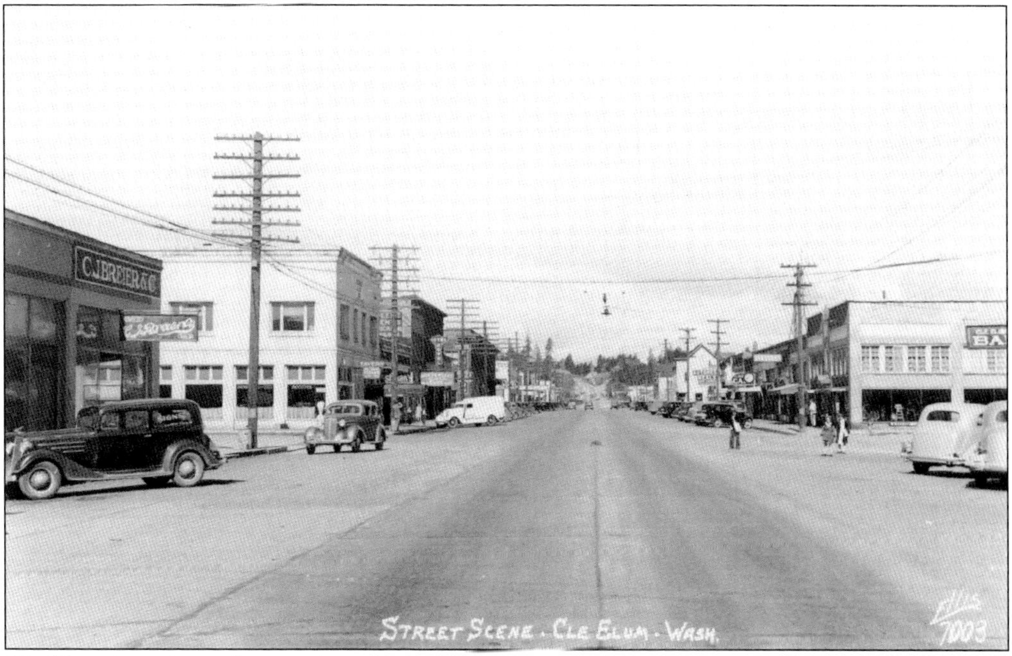

Cle Elum has always been a busy highway stop-over; as the last town of any size until North Bend, 50 miles west, it often finds itself hosting snowbound travelers when Snoqualmie Pass is closed by storms. This view looks west along 1st Street and shows the variety of businesses that catered to motorists as well as local ranchers and miners. The town was bypassed by Interstate 90 but remains vibrant.

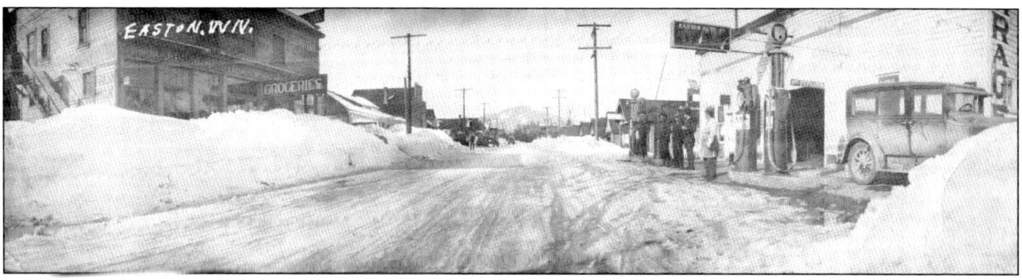

Easton was a railroad town. A 1909 insurance map shows a roundhouse, coal station, and businesses south of the tracks. When the highway was built through the north edge of town, stores and shops migrated to cater to the new mode of travel. Several buildings burned in 1934, including the Johnson Brothers store, but despite the devastating fire and being bypassed by Interstate 90, Easton hangs on. (KCHM, EA-02 98:51:40.)

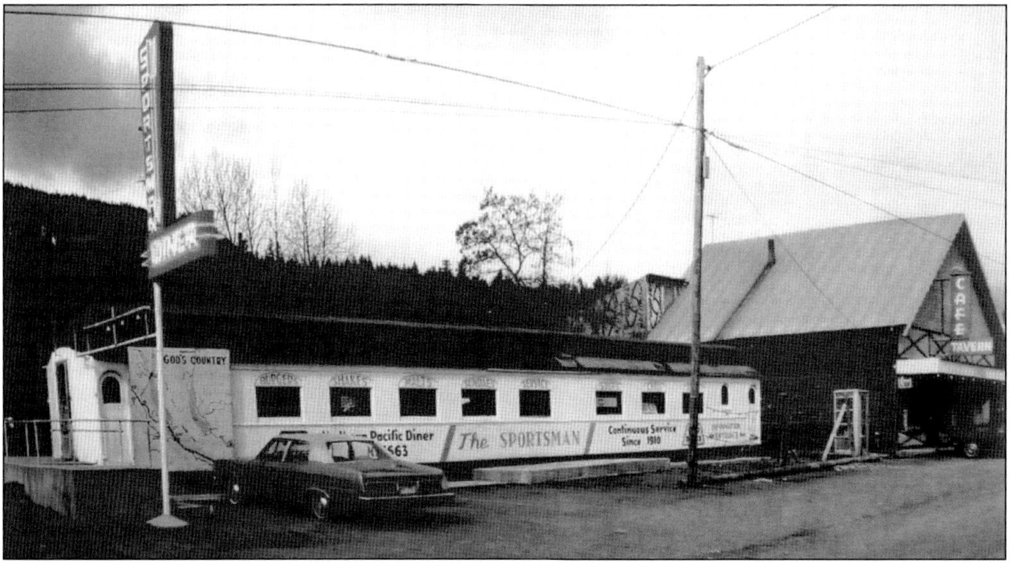

The Sportsman Diner was a long-time roadside fixture in Easton. Opened around 1950 by Edward "Lee" Peel, the Sportsman was housed in a former railroad dining car—Northern Pacific No. 1663, trucked in from Tacoma. A second railcar was used as a kitchen. The menu was typical diner fare. The diner closed in the 1970s, and car No. 1663 has been moved back to Tacoma for restoration.

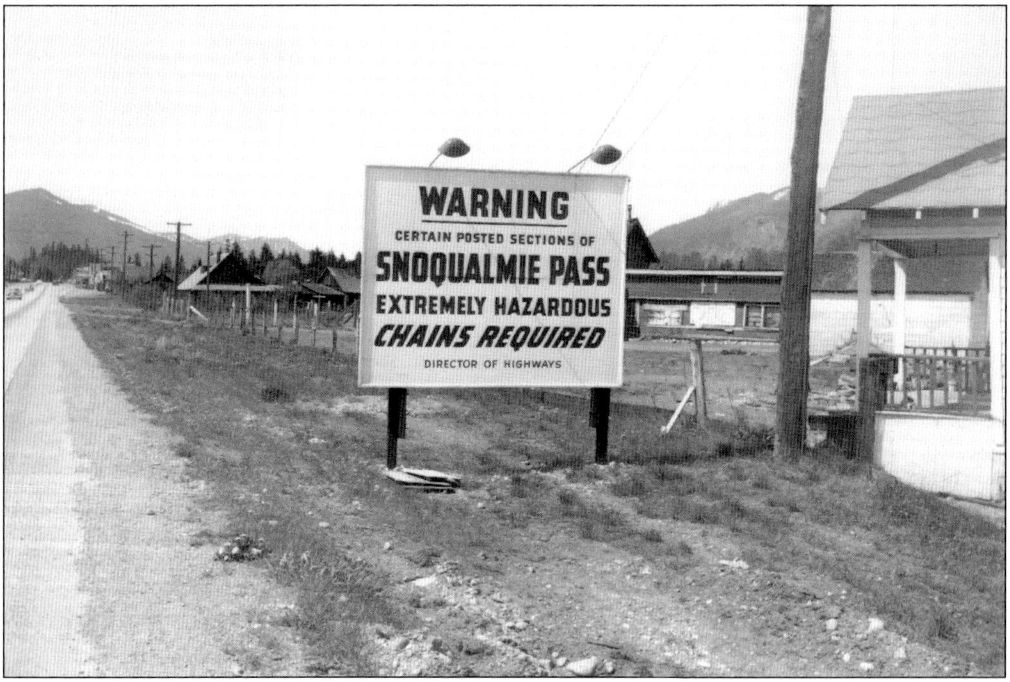

With the Cascade Mountains growing closer, westbound motorists began to see signs like this one outside of Easton. "Extremely hazardous" may have been an overstatement, but even as recently as 1947 when this photo was taken, winter mountain travel had its challenges—deep snow, avalanches, downed trees, and long distances between services—and was not to be taken lightly. (MOHAI, PI25291.)

Truckers and other frequent Sunset Highway travelers became accustomed to long delays while the road was cleared of snow and other obstacles. Even today the highway between Easton and North Bend is occasionally shut down for hours at a time due to extreme weather conditions. If the plows cannot keep up when heavy snowstorms hit, motorists can only wait it out. (MOHAI, PI25290.)

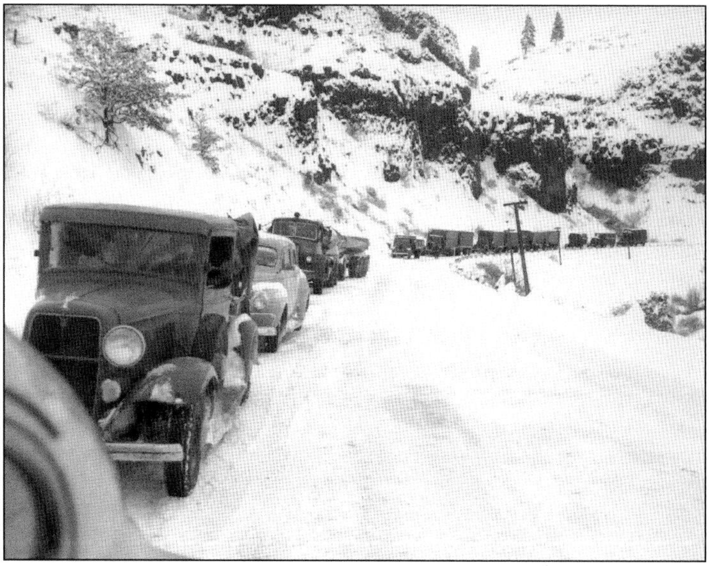

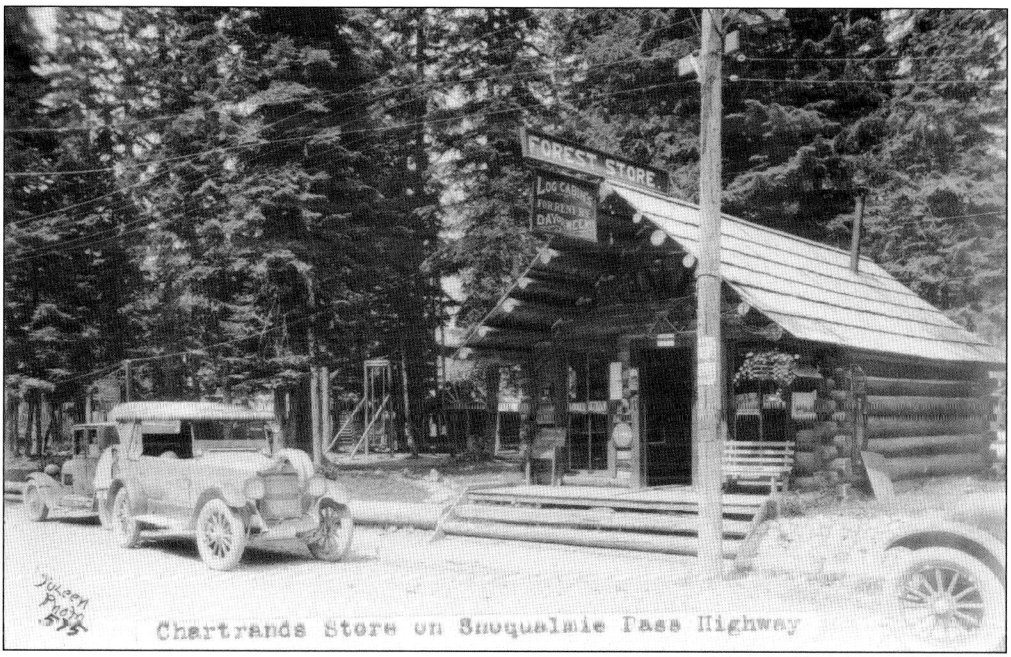

The Chartrand family homesteaded the upper Yakima River in 1911. A few years later, Ed Chartrand, taking note of the increasing numbers of travelers and campers on the nearby Sunset Highway, opened a campground and store 10 miles west of Easton at the turnoff to Lake Kachess. Chartrand's Grove and Forest Store catered to passing motorists as well as Seattleites, who made it a favorite camping spot. (EverettPL, J575.)

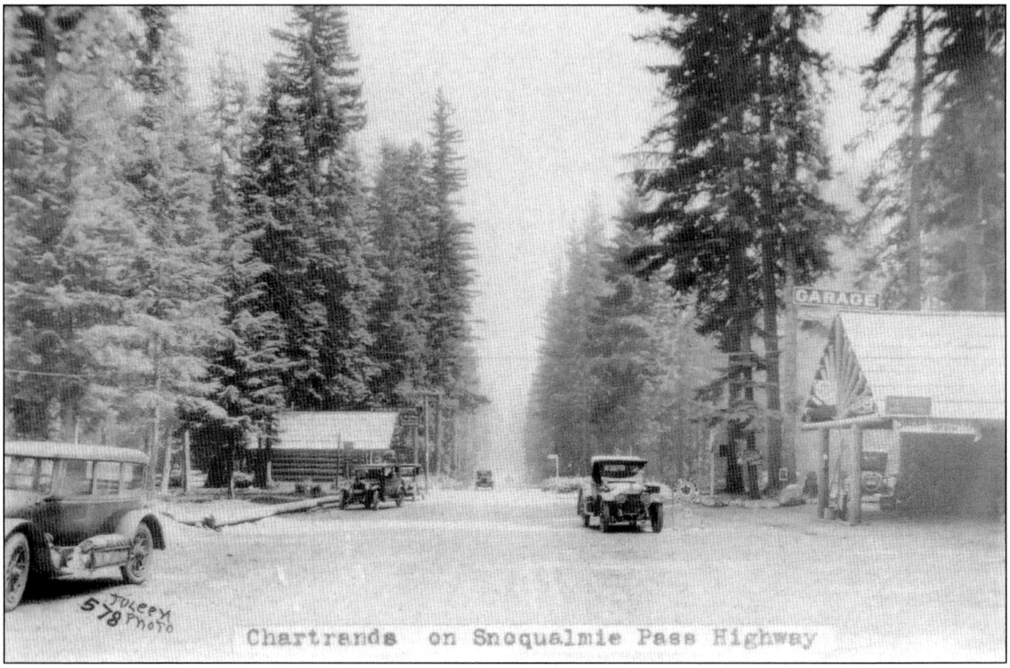

By 1925, several businesses had developed at the Lake Kachess junction. Chartrand's Log Cabins—with one, two, or three beds, stoves, and electric lights—had opened behind the Forest Store. The Speed Service Station—a single-pump garage selling Red Crown gasoline, tires, and Aristo motor oil—was built across the highway, and a new campground called Crystal Springs Camp had located next to the garage. (EverettPL, J578.)

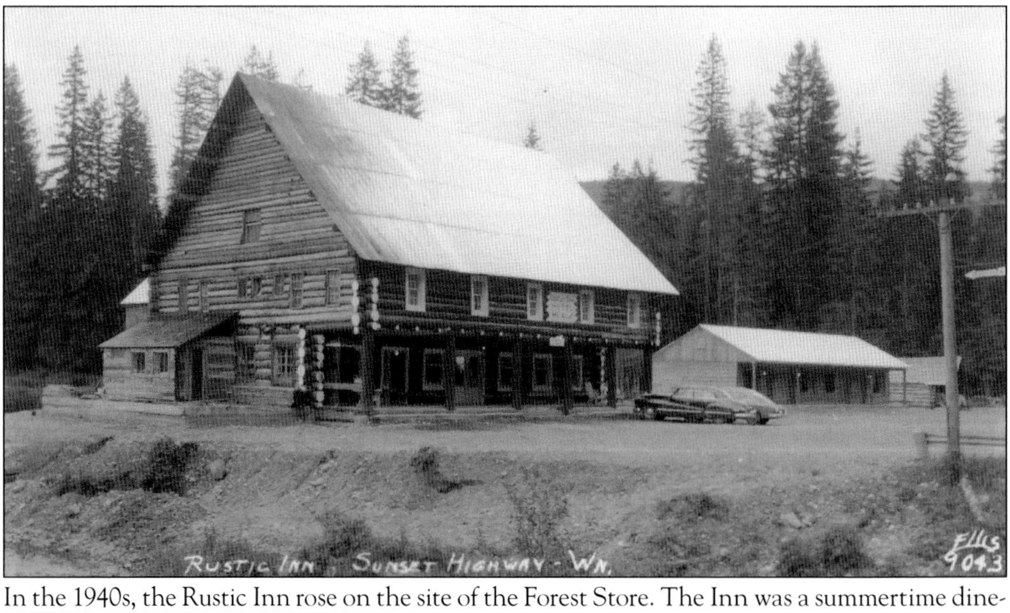

In the 1940s, the Rustic Inn rose on the site of the Forest Store. The Inn was a summertime dine-and-dance destination for Seattleites and a winter haven for skiers, operating its own toboggan and ski hill in the late 1930s. Downhillers at the nearby, now-vanished Stampede Pass ski run made the inn their ski lodge. Though condemned for highway expansion in 1957, the building survives as the Kachess Lodge.

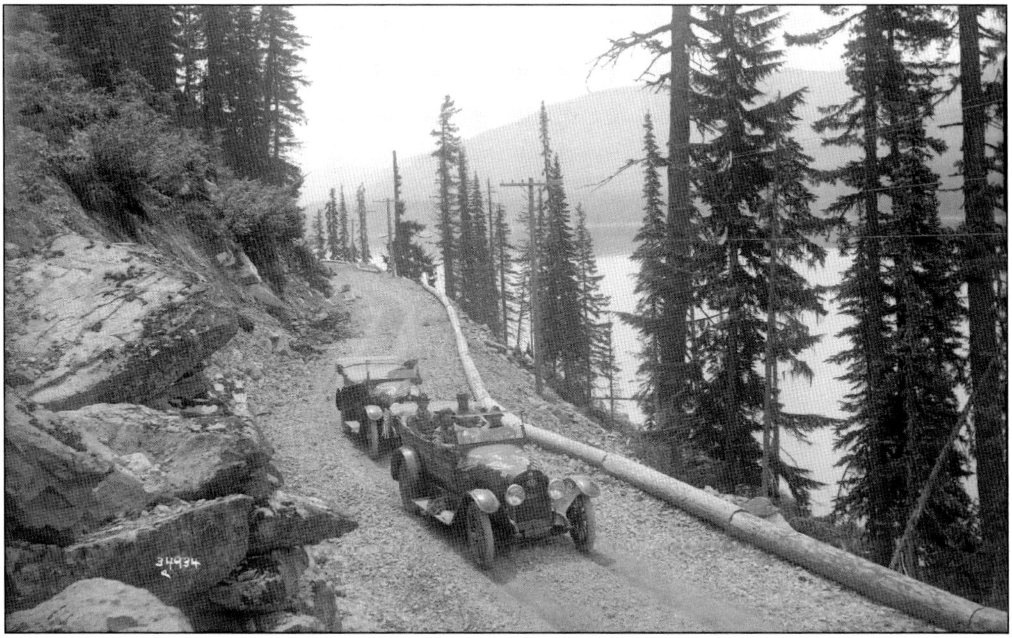

The National Park-to-Park Highway Association laid claim to the same route as the Sunset Highway over Snoqualmie Pass. In this view from 1916, association representatives motor along Lake Keechelus about five miles east of the pass. This segment of road, narrow and gravel-surfaced with logs laid as guardrails, was cut into the cliffside with a steep drop to the lake. The lead auto is identified as a Mitchell Six. (WSHS, 1943.42.34934.)

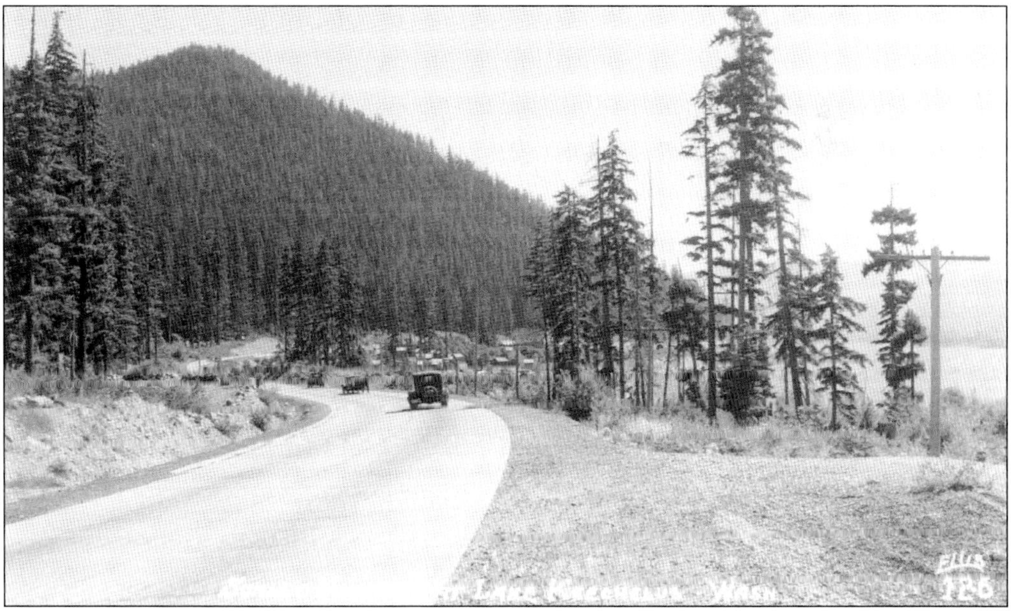

By contrast, the highway along Lake Keechelus had vastly improved a few years later. Travel across Snoqualmie Pass was becoming routine—no longer dreaded as in the early days of muddy, rutted, snow-bound roads. Two lanes of smooth concrete with wide gravel shoulders reflect modern construction, though guard rails are conspicuously absent. A cluster of vacation cabins is seen around the curve in the road.

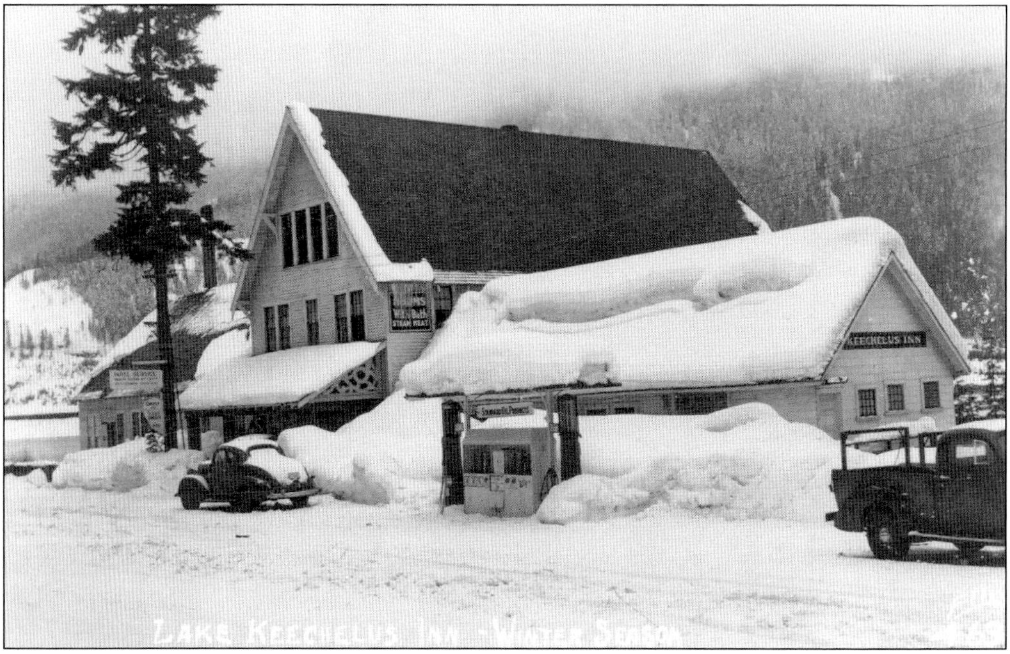

As early as 1914, the Keechelus Inn advertised excursions from Seattle via the Milwaukee Railroad. Access improved when the Sunset Highway was laid out over the old wagon road across Snoqualmie Pass. Situated on Lake Keechelus east of the pass, the inn, a "resort of excellent appointments," offered hotel and tent accommodations. The dining room served chicken and trout dinners, and a store and service station were nearby.

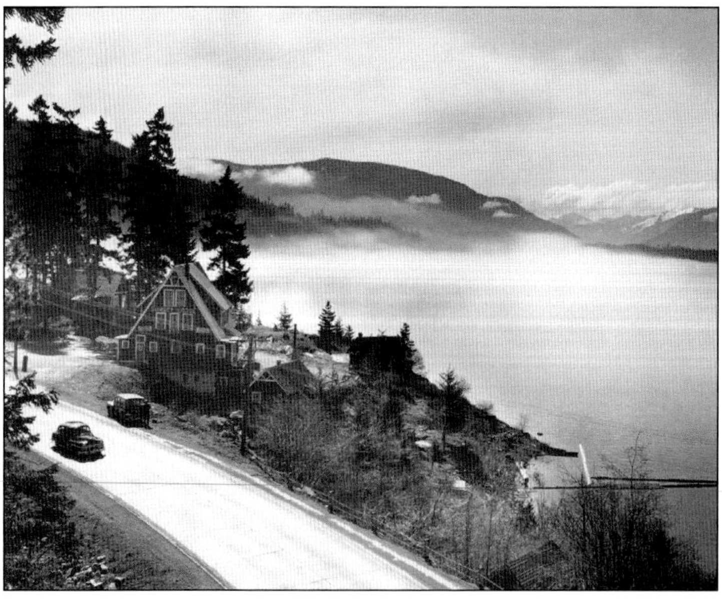

The Sunset Lodge, a mile west of Keechelus Inn, celebrated its grand opening with a grand ball on July 7, 1917. The Lodge, "A Homelike Summer Resort," was managed by Leslie W. Hayes and offered hotel rooms, cottages, and fully equipped "camping out" tents. À la carte meals were available at all hours. Sunset Highway's expansion to four lanes doomed the Lodge in 1956. (WSDA, AR-28001001-ph001159.)

By 1867, a primitive road crossed Snoqualmie Pass, linking the wet and dry sides of the territory (Washington did not become a state until 1889). In 1883, the Seattle and Walla Walla Trail and Wagon Road Company assumed ownership and financed improvements by charging tolls. Portions of the wagon road, such as here along Lake Keechelus, became State Road 7, later incorporated into the Sunset Highway in 1913. (UW, TRA0220.)

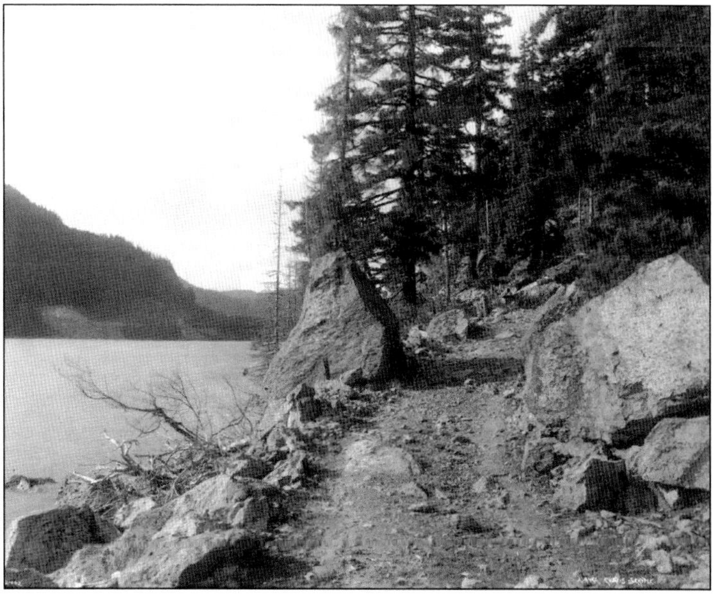

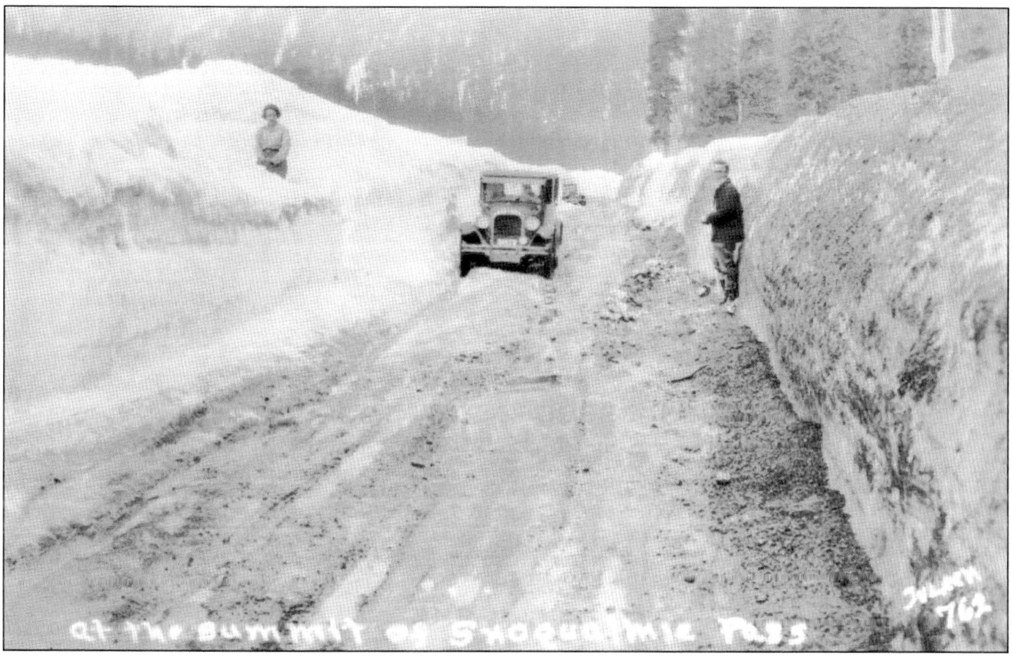

Snoqualmie Pass was officially closed to travel each winter until 1931, though some people braved the dual threats of weather and officialdom to chance the crossing. Even after the seasonal opening, usually in May, six-foot snow banks were common along the highway. This couple left their car parked right on the highway while they had fun in the snow. There was so little traffic that probably nobody noticed. (MOHAI, 1998.10.14.)

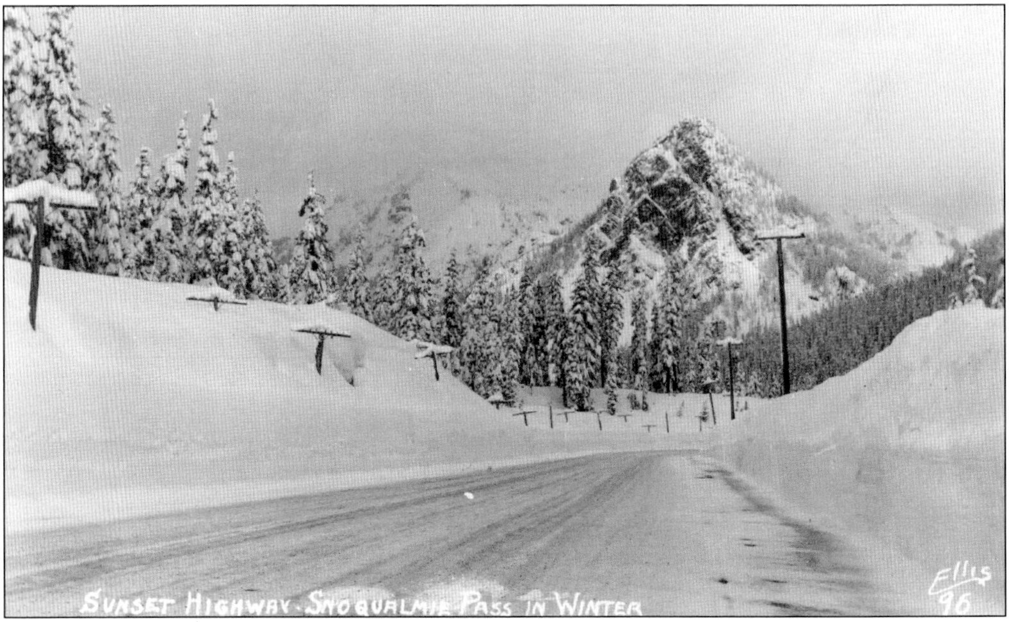

The Department of Highways' October 1930 decision to keep the highway open all year was hailed by newspapers on both sides of the pass, though some complained the light traffic could not justify the expense. The objections proved wrong—traffic now averages 28,000 vehicles daily, though scarce the day this photo was taken in the 1950s. The view looks west toward Mount Snoqualmie and Guye Peak. (SVM, PO.040.3186.)

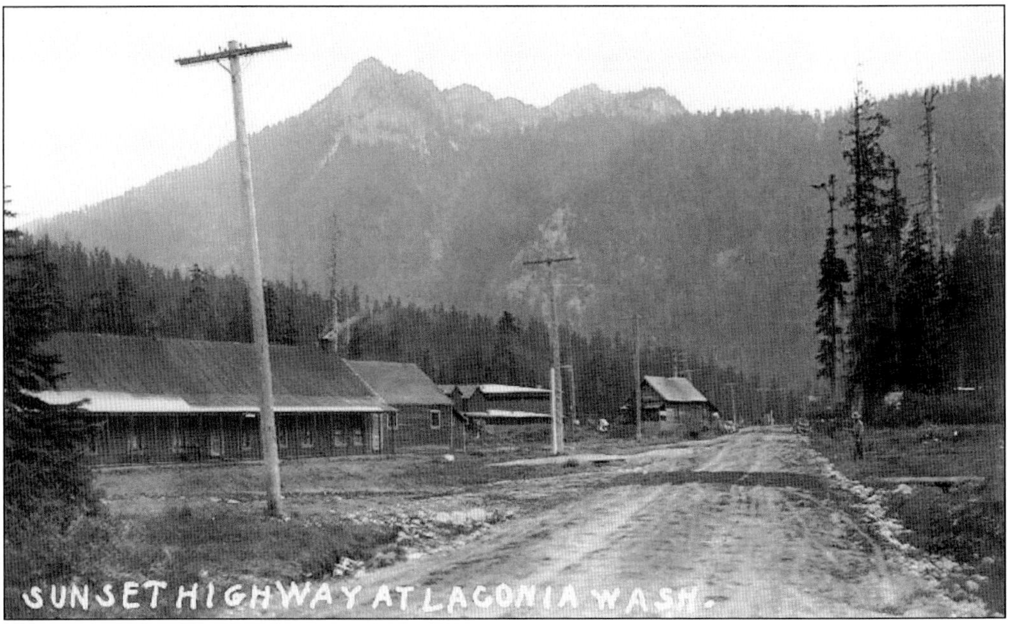

When the Chicago, Milwaukee & Puget Sound Railway laid rails over Snoqualmie Pass in 1908, a collection of railroad structures—a depot, an engine house, a turntable, a water tank, a hotel, and a bunkhouse—was established at the summit and named Laconia. In 1915, the railroad opened a tunnel underneath the pass; no longer servicing trains, Laconia catered to early Sunset Highway travelers. The hotel, on the left, burned around 1920. (SVM, PO.076.0050.)

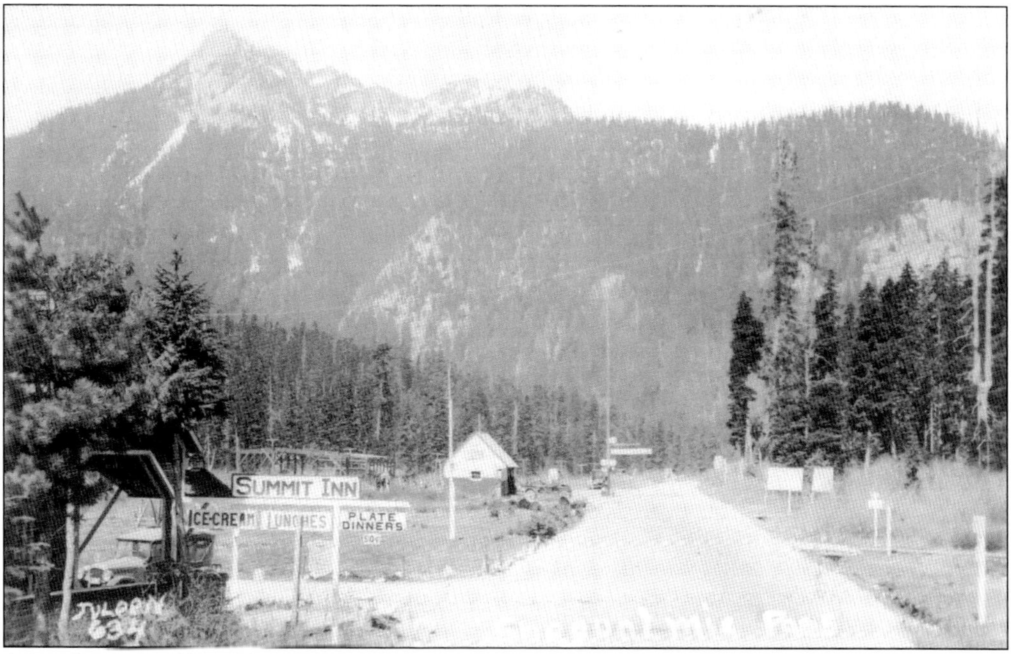

After the original hotel burned and Laconia had ceased to function as a railroad town, Paul Holden bought the old Milwaukee depot and converted it into the Summit Inn. This view shows the inn in its original position on the south side of the road. In the distance is the Summit Garage. Just beyond the garage, the highway begins its long descent into the Snoqualmie River valley.

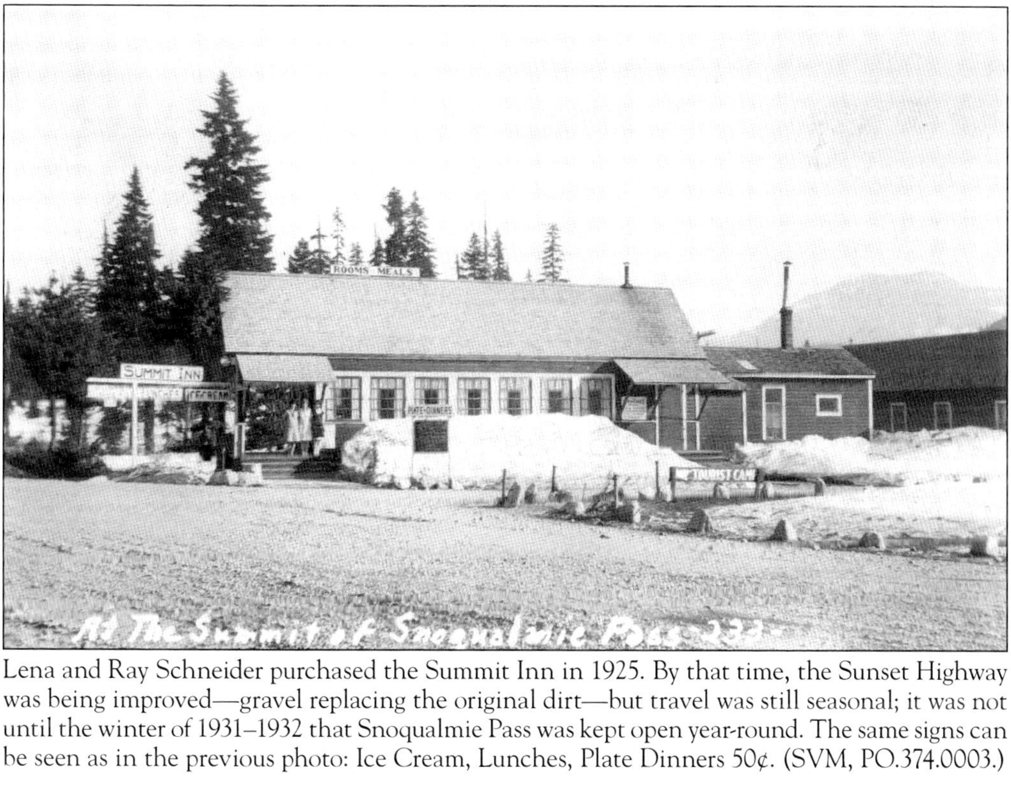

Lena and Ray Schneider purchased the Summit Inn in 1925. By that time, the Sunset Highway was being improved—gravel replacing the original dirt—but travel was still seasonal; it was not until the winter of 1931–1932 that Snoqualmie Pass was kept open year-round. The same signs can be seen as in the previous photo: Ice Cream, Lunches, Plate Dinners 50¢. (SVM, PO.374.0003.)

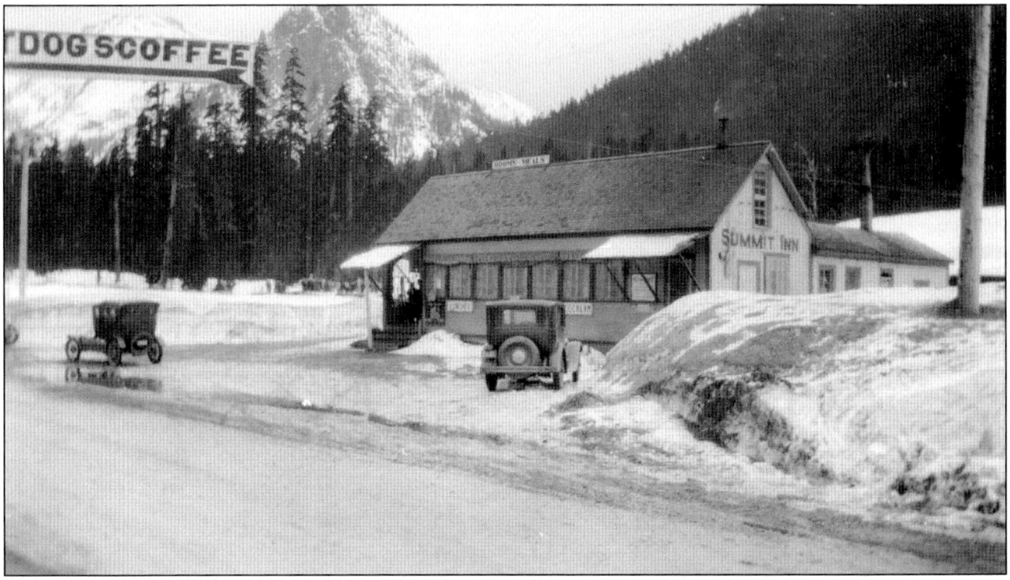

Shortly after the Sunset Highway became a year-round route across Snoqualmie Pass, the Schneiders moved the Summit Inn across to the north side of the highway and turned it to face the road. An addition was constructed to the repositioned inn with eight bedrooms, each with double bed, linen, blankets, washbowl, and pitcher (no running water, no sink), and chamber pot (with optional outside toilets). (SVM, PO.374.0016.)

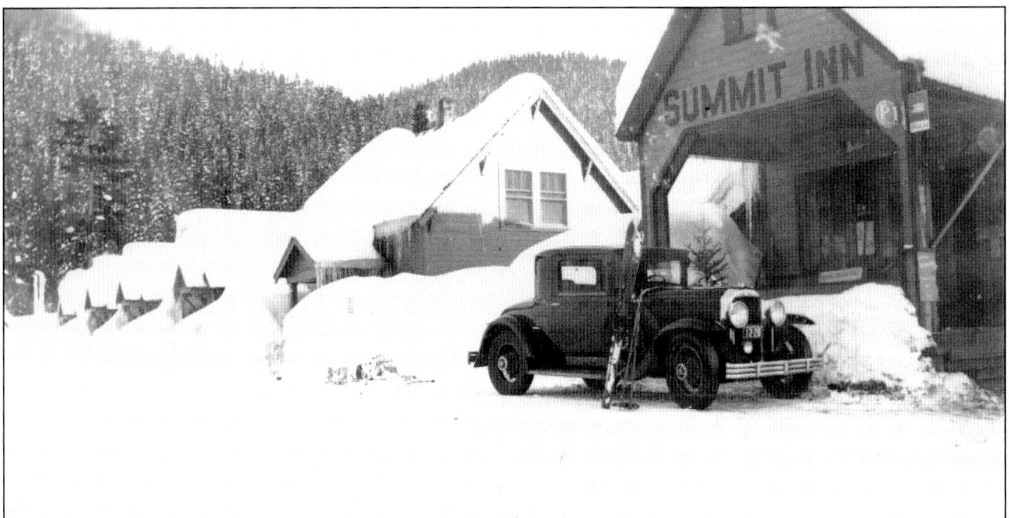

In 1931, the Schneiders added four tourist cabins behind the Summit Inn, replacing an earlier tourist campground at the rear of the building. The Schneiders' house was immediately to the left of the entrance to the inn, with the tourist cabins farther along the road. A pair of skis rest against the old auto in this wintry scene. (SVM, PO.374.0032.)

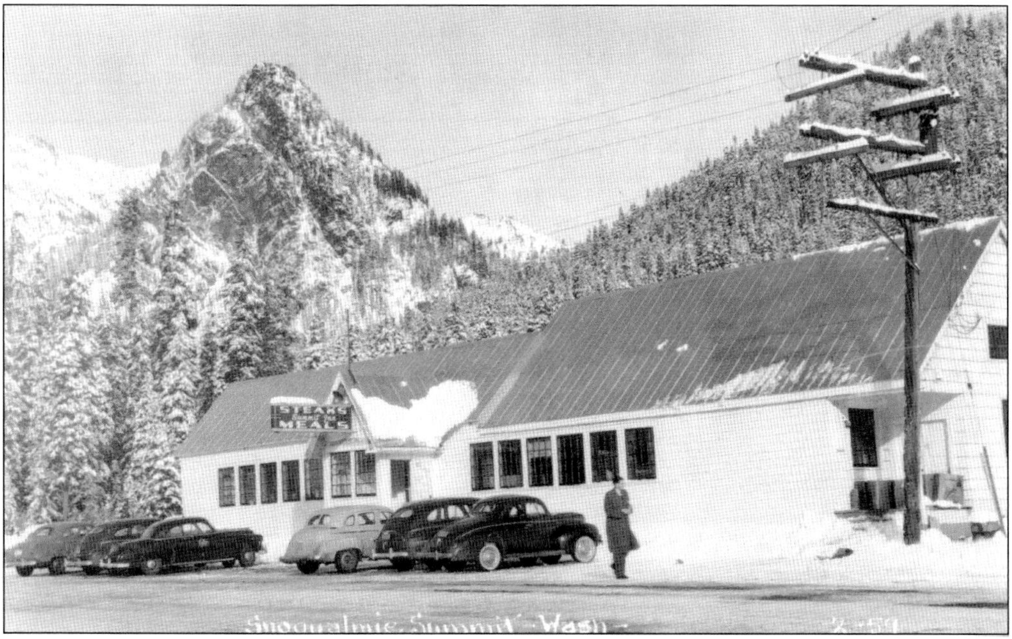

Marshall C. "Cruse" Mason purchased the Summit Inn from the Schneiders in the 1930s. Mason was also associated with Camp Mason a few miles farther west on the Sunset Highway. By 1949, the Summit Inn was being operated by Randall Green. The inn continued to expand over time, with a gable built over a new entrance and a large addition constructed on the east end of the original building.

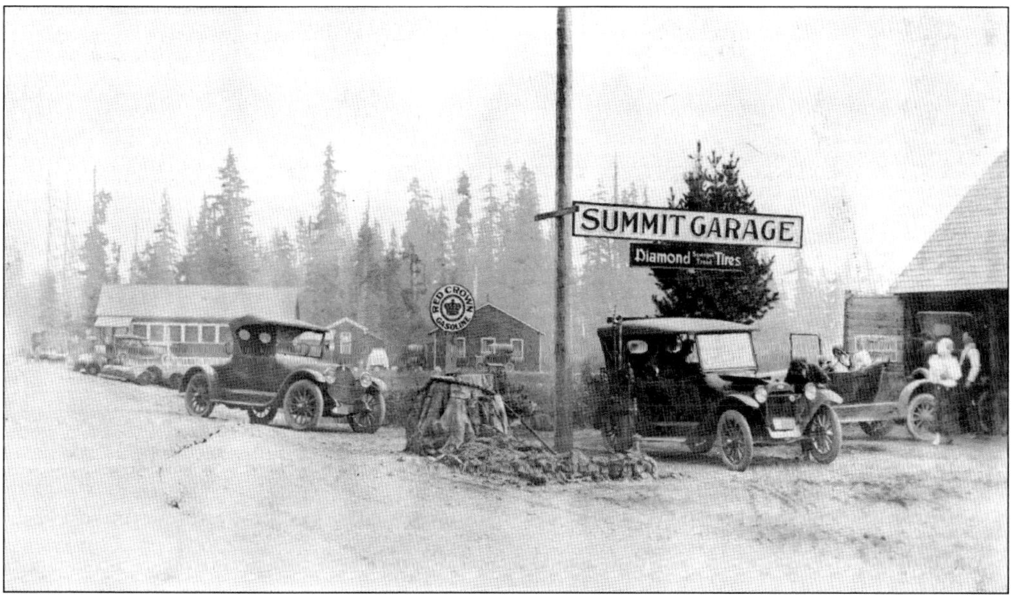

Dating from at least 1920, the Summit Garage was west of the original Summit Inn, seen in the background. A simple two-door shack, the garage provided towing service, Diamond Tires, and Red Crown gasoline dispensed from the crude-looking pump next to the telephone pole. The garage updated its appearance over the years—a more substantial building appeared in the 1940s—but has vanished. (SVM, PO-393-2.)

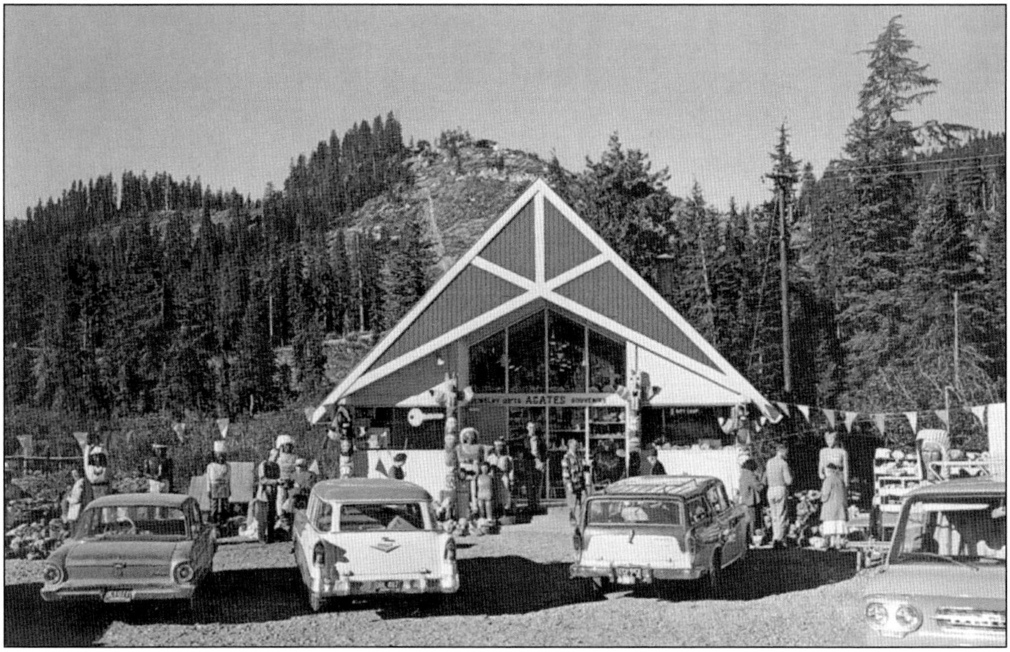

The Rock Chalet called itself "one of the most unusual and complete" gift shops in the country. Featured at the shop were large, hand-carved wooden Indians created by Gene Delaney, an Indian living in Oklahoma. As many as 50 or 60 carvings were on display at a time. The Chalet was operated by Jack Preston in the 1960s. Only its foundation remains, across from the Summit Inn. (JCO'D.)

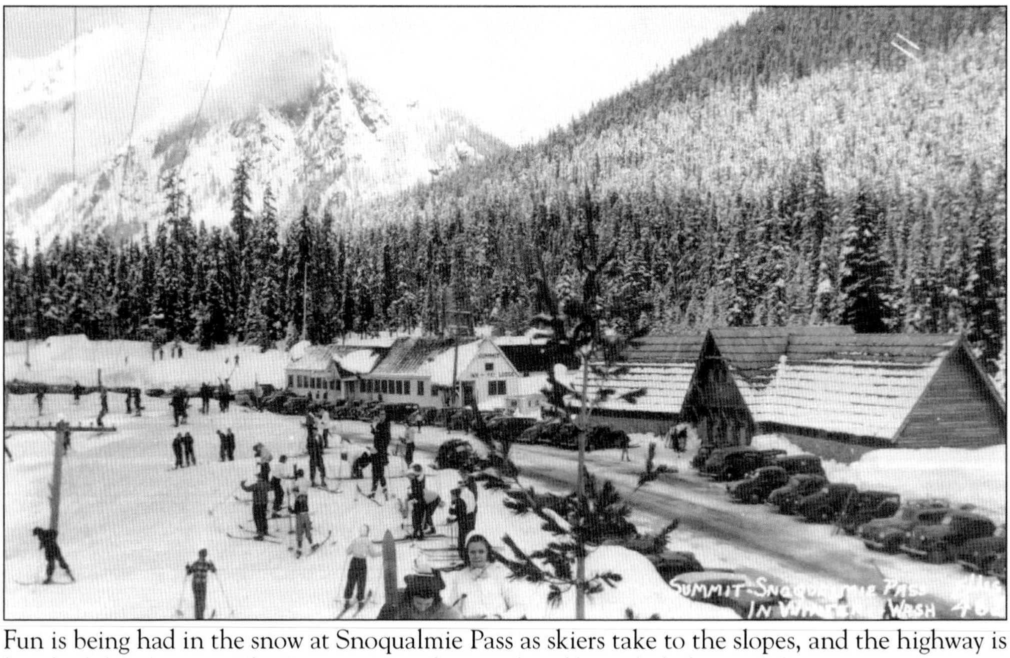

Fun is being had in the snow at Snoqualmie Pass as skiers take to the slopes, and the highway is lined solid with autos. The Summit Inn advertised itself as a ski lodge as early as 1932; in winter it stayed open 24 hours a day to handle the crowds. The inn burned in October 1968. The comfort station next door, called Travelers' Rest, was a 1938 WPA project and is still in use.

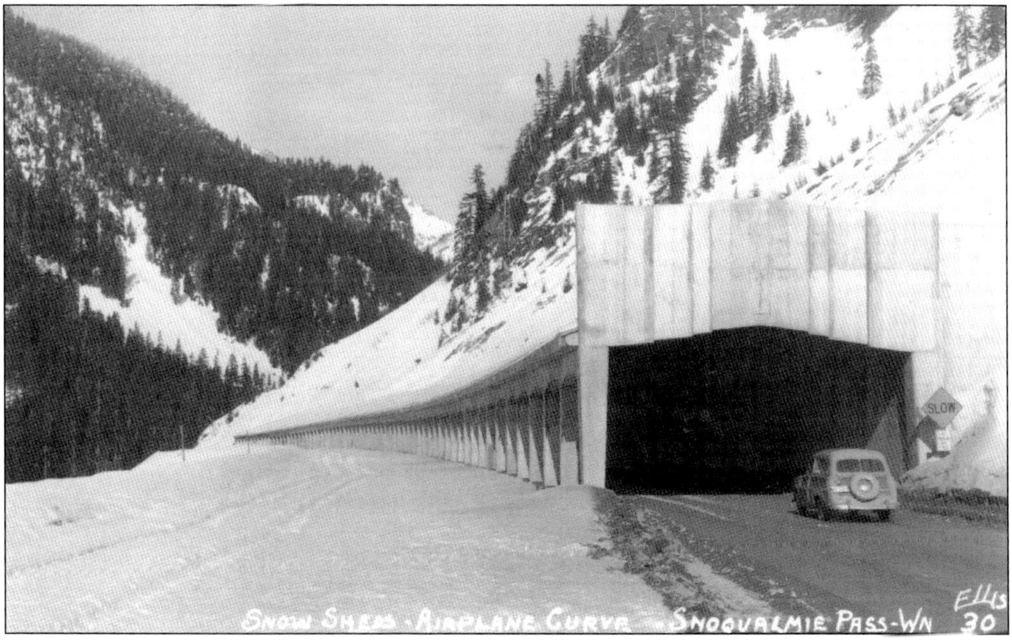

Airplane Curve, a mile and a half west of the summit, received its name when, in November 1931, an airplane carrying five hunters crashed into the rock bluff immediately above the highway and tumbled onto the road. The location of numerous avalanches over the years, a 1,300-foot-long snowshed was finally constructed in 1950. The highway was expanded to four lanes soon afterwards. The concrete snowshed is now gone.

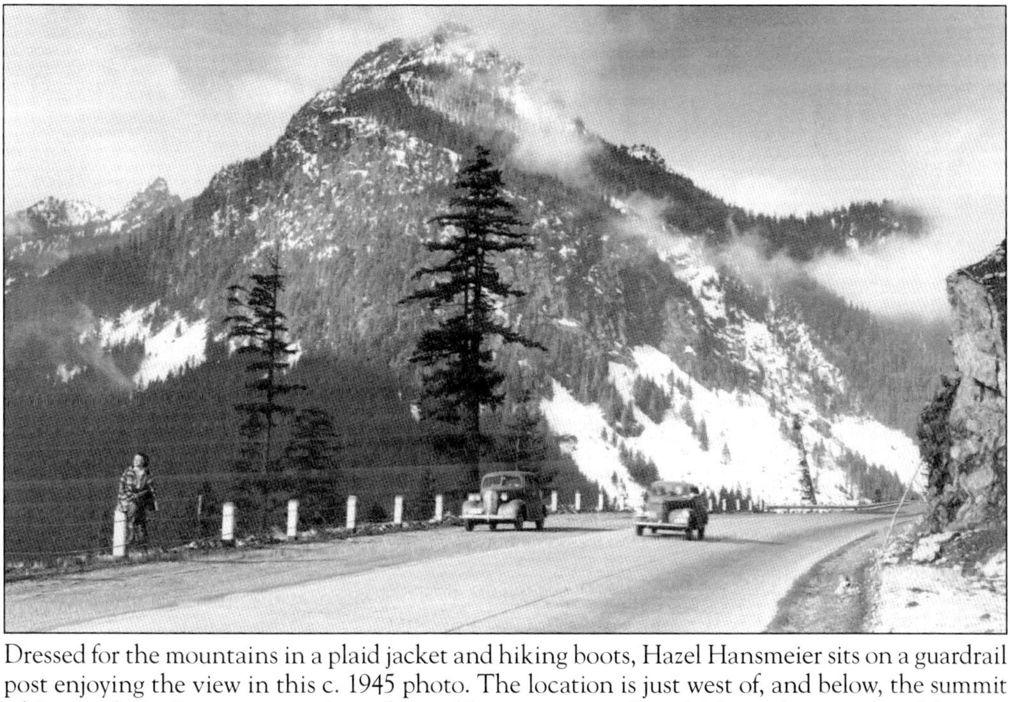

Dressed for the mountains in a plaid jacket and hiking boots, Hazel Hansmeier sits on a guardrail post enjoying the view in this c. 1945 photo. The location is just west of, and below, the summit of Snoqualmie Pass, on a portion of the old concrete two-lane highway later replaced by new construction. Denny Mountain forms the background. (SVM, PO.304.0010.)

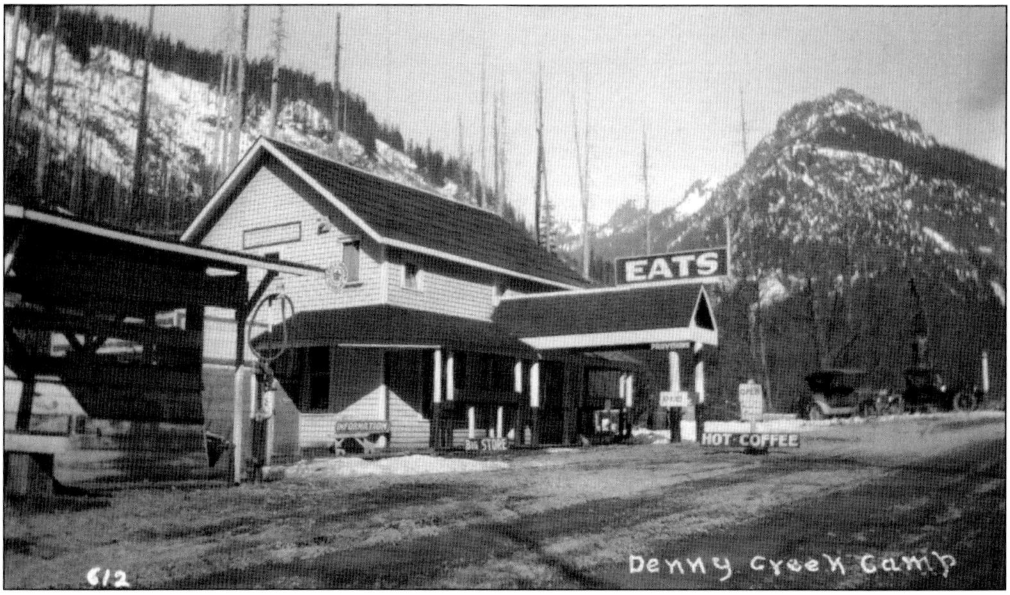

Denny Creek Camp is situated at the west base of Snoqualmie Pass. This 1925 view shows the service station and general store that was located where the road to the campground turned off the old Sunset Highway. In addition to Red Crown gasoline, signs advertising hot coffee, pie, and provisions can be seen against the backdrop of the rugged Cascade Mountains. (EverettPL, J612.)

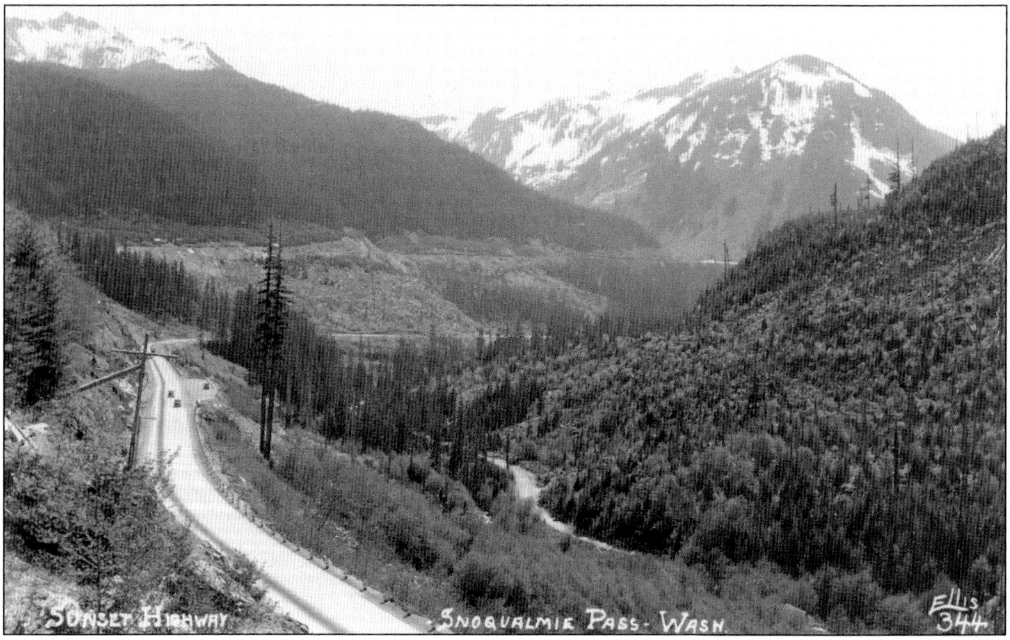

Heading west from Snoqualmie Pass, the Sunset Highway paralleled the Snoqualmie River—barely more than a narrow stream at this point—through the Cascade Mountain foothills. Portions of the original highway still exist, bounded on both sides by six-lane Interstate 90. Today there are virtually no services along the 20-mile segment of interstate highway between Snoqualmie Pass and North Bend.

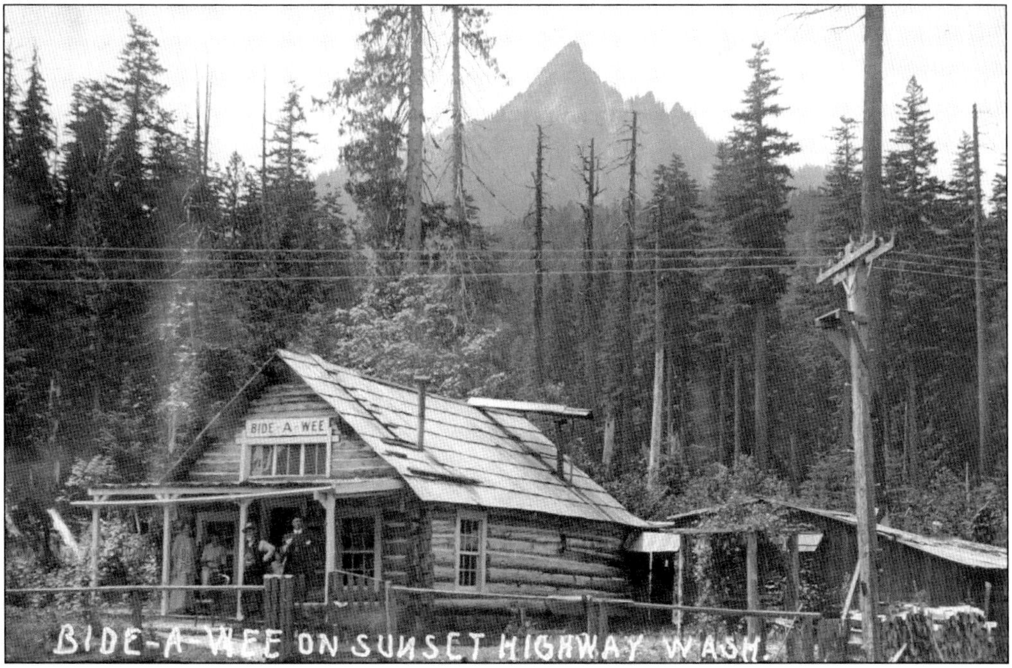

Bide-a-Wee Ranch, an early travelers' stop on the Snoqualmie Pass wagon road, was located about 10 miles west of the pass. Railroad construction over the pass in 1909 diminished Bide-a-Wee's importance as a way station, but the opening of the Sunset Highway and the growing popularity of automobile driving reversed its fortunes—it became a tourist camp and popular starting point for Cascade Mountains hikes. (MOHAI, 1998.10.12.)

M.C. Mason purchased Bide-a-Wee in 1919, renamed it after himself, and upgraded the premises. Instead of the earlier primitive facilities, Camp Mason provided 25 cottage cabins having two beds, a cook stove, cooking utensils and dishes, and available with or without bedding. A lunchroom, grocery store, and gas station were also on the premises. Offering weekly and overnight accommodations, Camp Mason appealed to both vacationers and Sunset Highway travelers. (UW, LIN0133.)

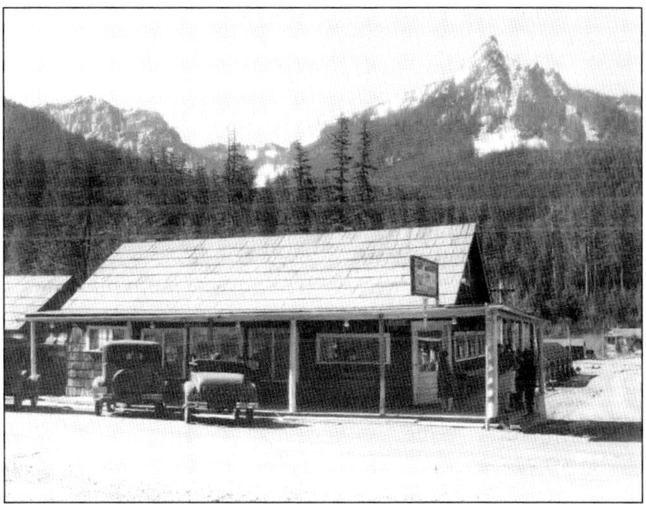

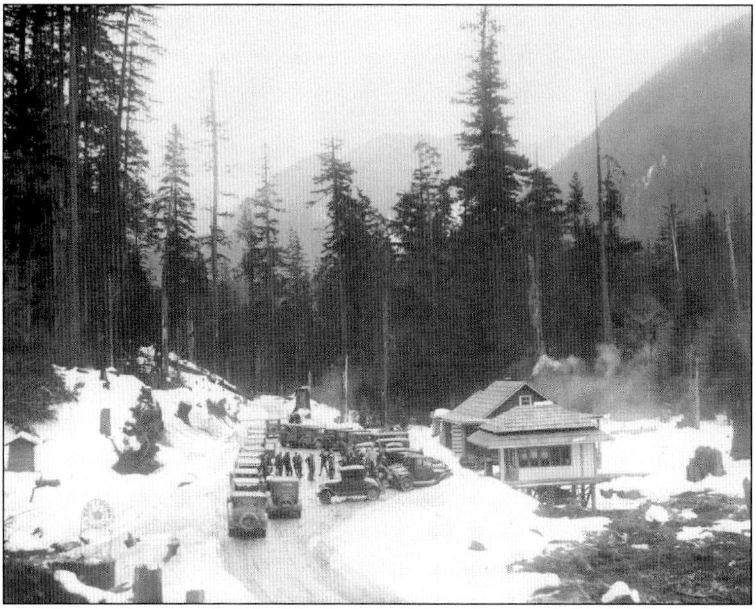

For many years, Camp Mason was the end of the road for eastbound winter travel. It was not until the winter of 1930–1931 that the Sunset Highway was kept open year-round. By 1949, Camp Mason was being operated by E.L. Munson, and in 1952, the restaurant was owned by Mr. and Mrs. Russell Creighton. The site is now buried under Interstate 90. (MOHAI, PI25253.)

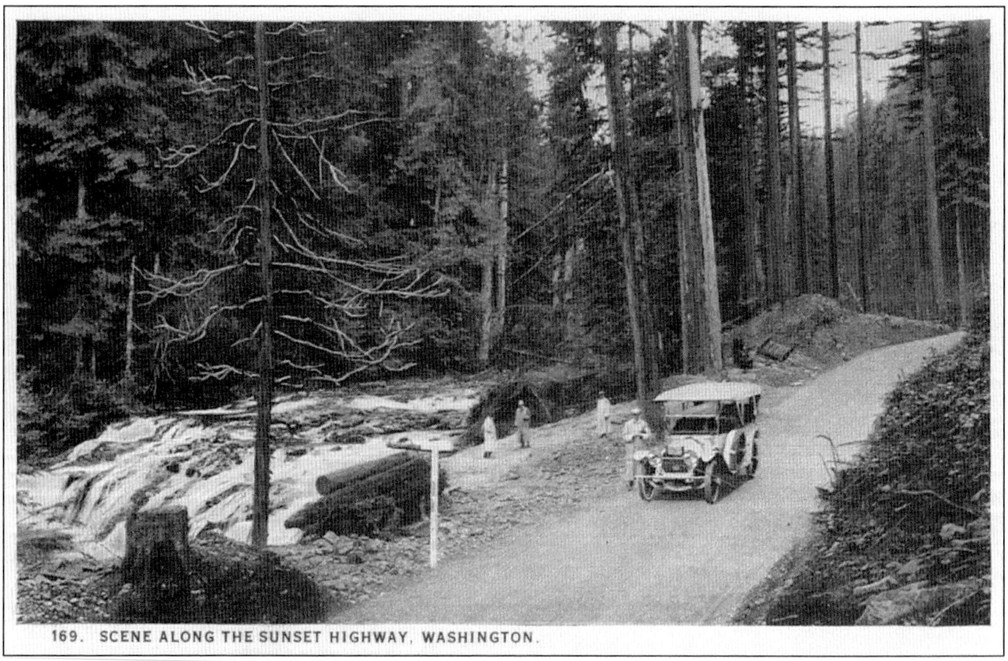

With tall trees and waterfalls, the Sunset Highway wound its way through a scenic wonderland as it followed the Snoqualmie River. This view shows a group of motorists taking in the sights at Weeks Falls, one of five waterfalls between the pass and North Bend. A segment of abandoned highway provides access from Interstate 90 to Weeks Falls, located in Olallie State Park.

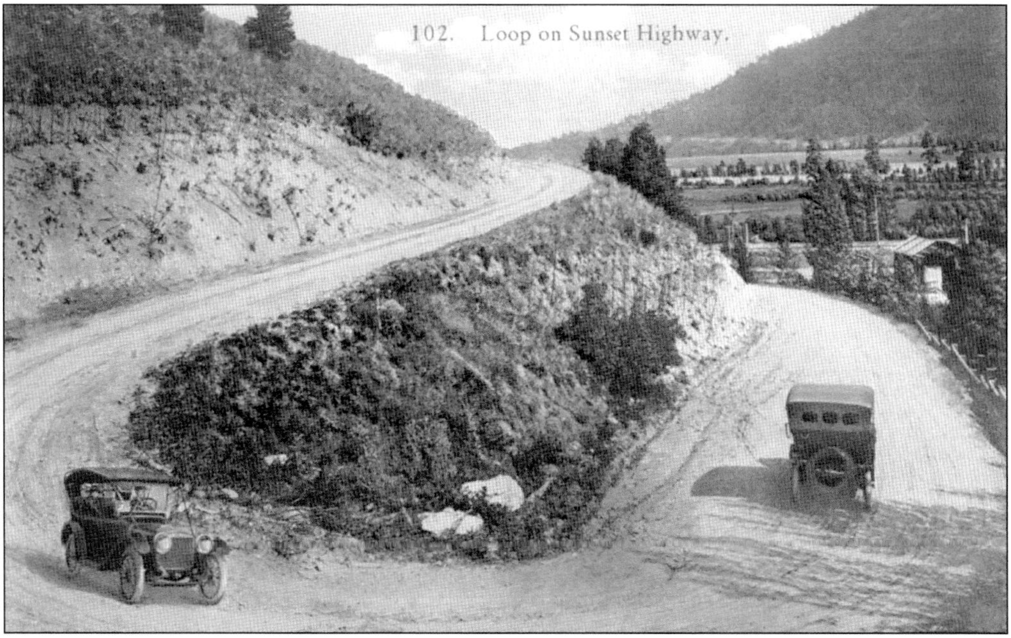

As it exited the Cascade Mountains, the highway dropped 400 feet in less than two miles to emerge onto the broad floodplain of the Snoqualmie River. In early times, the elevation drop was accommodated by a series of hairpin turns, as this view shows. Today, Interstate 90 makes the grade change in a single swoop of modern concrete pavement; all traces of the old highway are long gone.

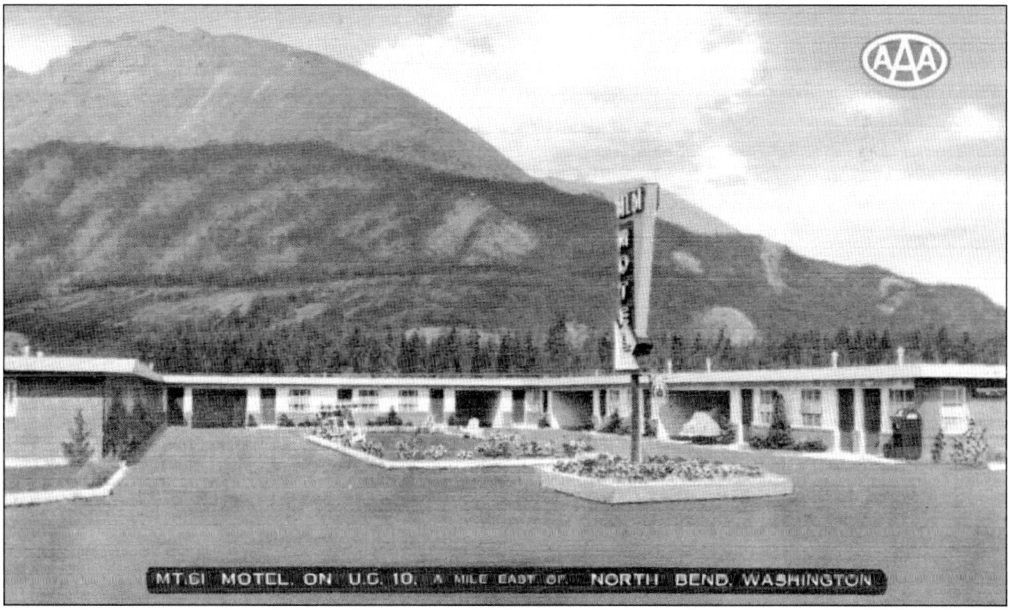

In 1947, the 17-unit Mt. Si Motel, a mile east of North Bend, was managed by O. Rainwater. The layout was the inverted-U shape design popular at the time, with ornamental plantings and lawn chairs on a center lawn. A towering neon sign displaying the motel's AAA recommendation edged the highway. The Mt. Si appeared in the cult classic *Twin Peaks: Fire Walk With Me* as the Blue Diamond Motel.

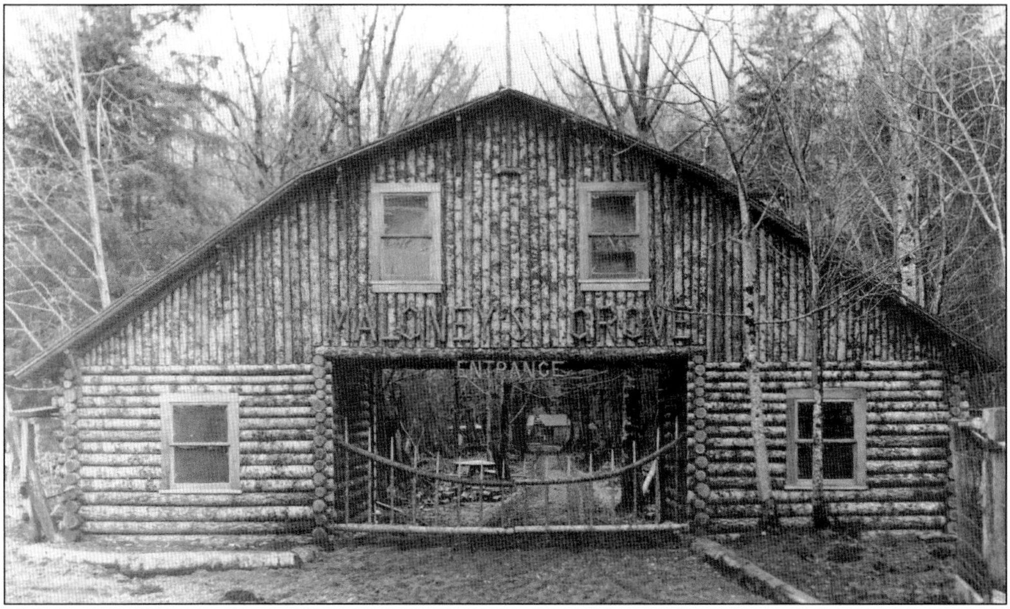

In 1913, Peter Maloney turned part of his farm east of North Bend into a picnic grounds and camping resort called Maloney's Grove. One of the earliest auto camps in the Northwest, the grove sat in a shady grove on the Snoqualmie River and offered 21 furnished cottages, each with two beds, a stove, dishes, electric lights, and firewood for between 50¢ and $2 per day. (SVM, PO.308.0178.)

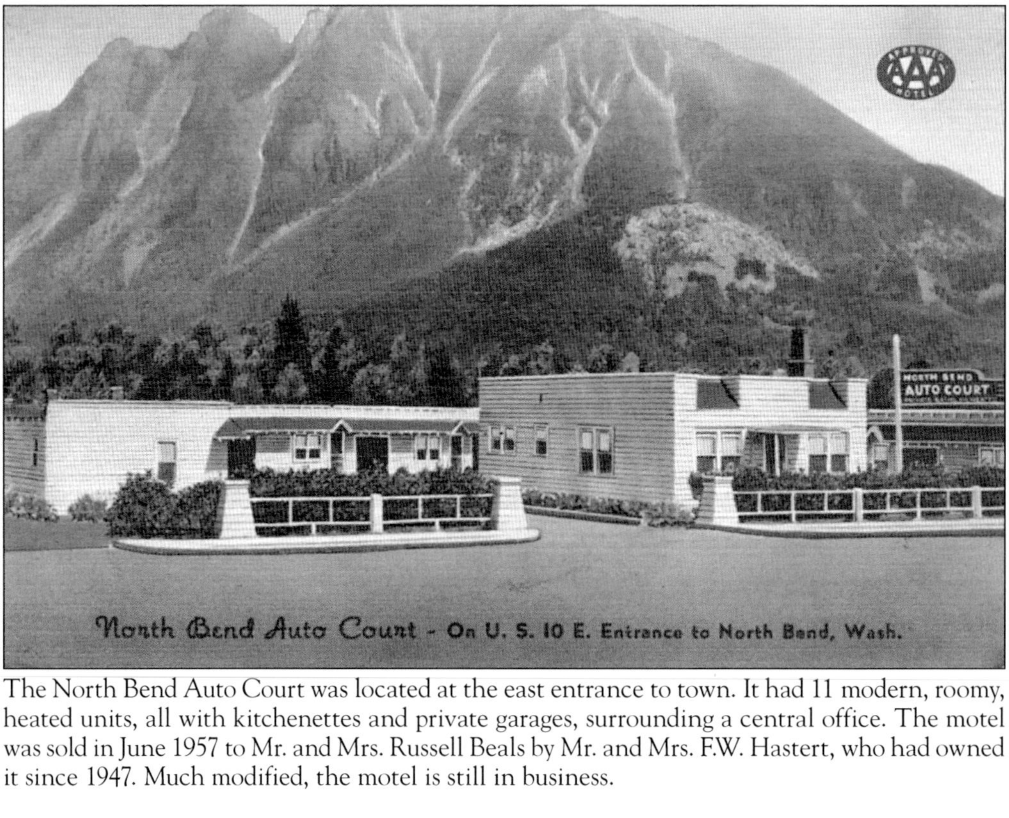

The North Bend Auto Court was located at the east entrance to town. It had 11 modern, roomy, heated units, all with kitchenettes and private garages, surrounding a central office. The motel was sold in June 1957 to Mr. and Mrs. Russell Beals by Mr. and Mrs. F.W. Hastert, who had owned it since 1947. Much modified, the motel is still in business.

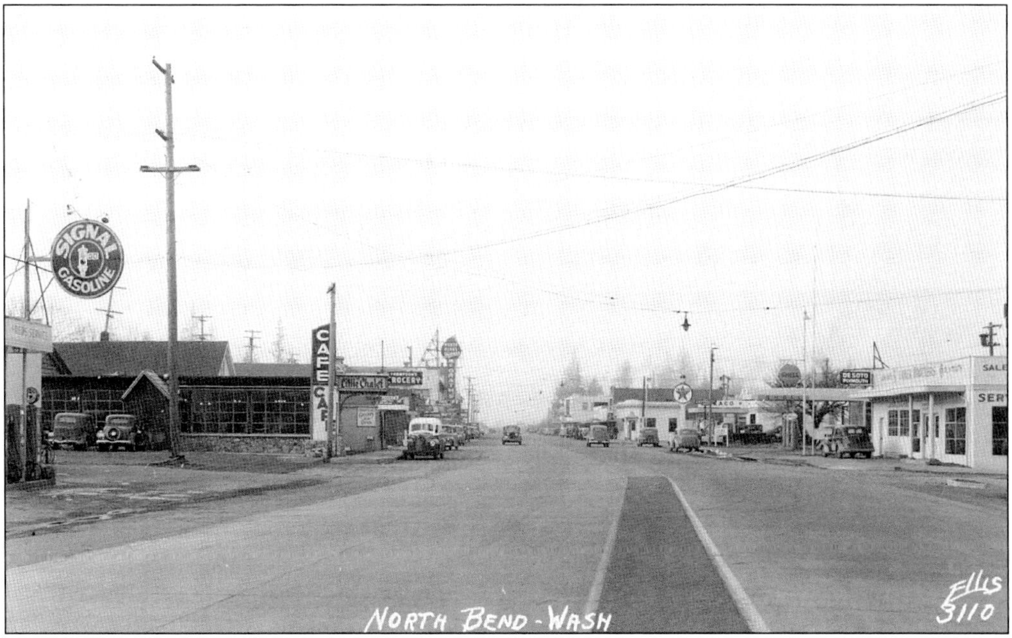

North Bend was the Sunset Highway traveler's gateway to western Washington. The approach to town was heralded by a plethora of roadside businesses, as this photograph looking west along the divided highway shows. Visible here are signs for Shell and Texaco service stations, the Little Chalet Café, Thompson's Grocery, and McGrath's Hotel, "North Bend's Finest." Another hotel, the McClelland, is visible in the far distance.

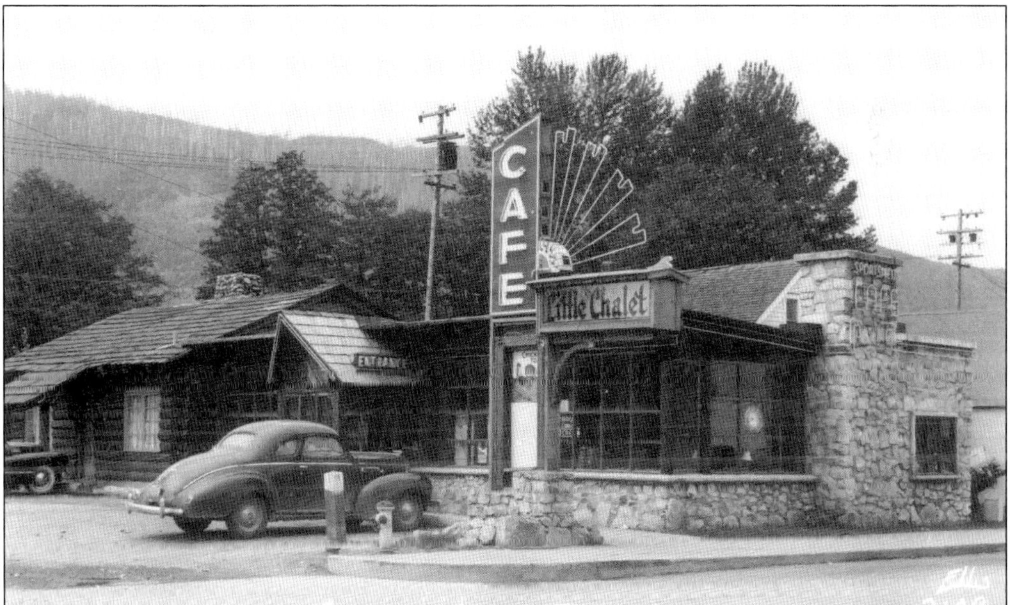

The Little Chalet Café, at 145 E North Bend Way, was open for breakfast, lunch, and dinner. Originally a log cabin with a stone-faced front addition, by 1949, the café had adopted a more updated design and was known as the Little Chalet Pancake Shop. A major facelift in 2008 obscured much of the original building, though it is still in business as the North Bend Bar & Grill.

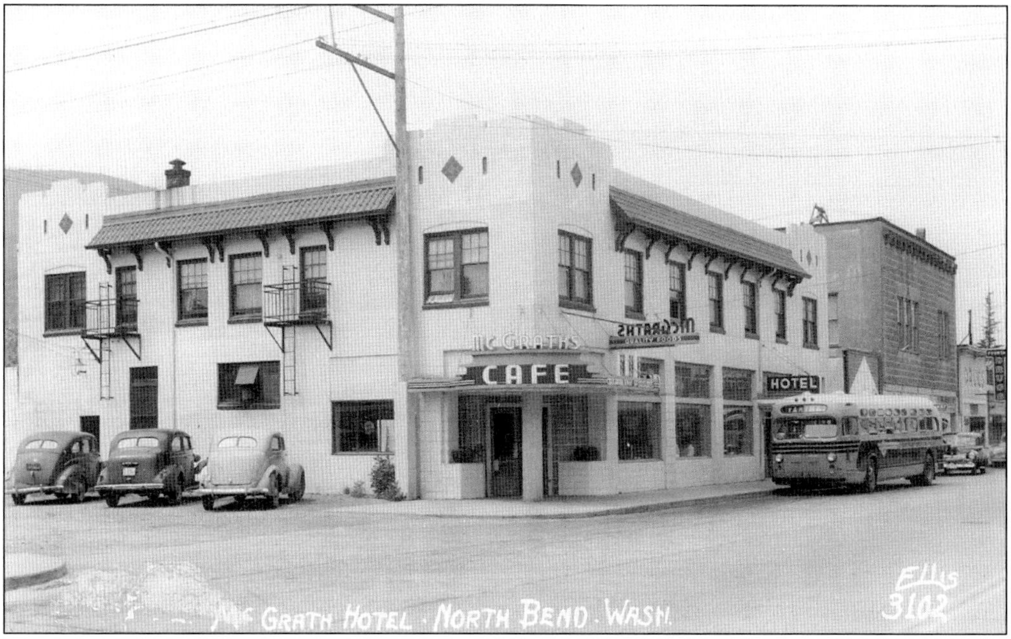

The McGrath Hotel was a family affair: Jack operated it with brother Harry, and sister Mary worked the café. Opened around 1920 as a single-story café, by 1926, a second floor of hotel rooms had been added, and the first floor expanded to include a lobby. The café offered Sunday dinners for 50¢ and 75¢. After Jack McGrath sold the business in 1944, it became the Thompson Hotel and Café.

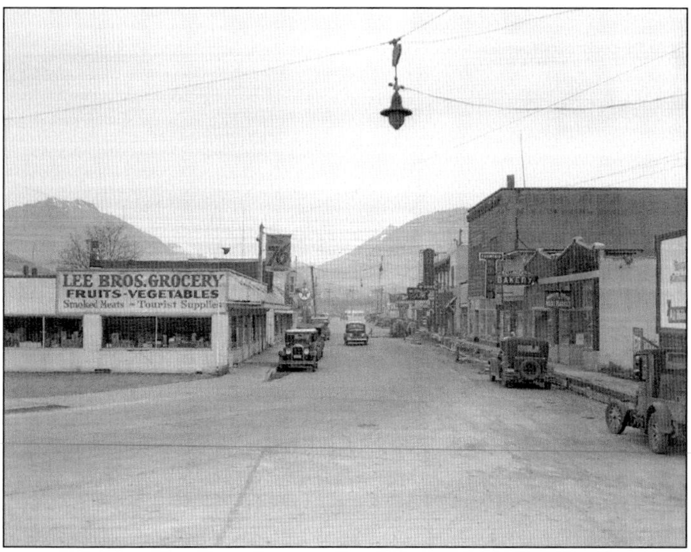

The early Sunset Highway through North Bend traversed a narrow main street, as shown in this view looking east. Although the business center only ran the length of three blocks, heavy highway traffic combined with local residents patronizing downtown stores combined to cause frequent congestion. Signs for Lee Bros. Grocery and a 76 service station can be seen on the north side of the block. (MOHAI, 1986.5.5984.1.)

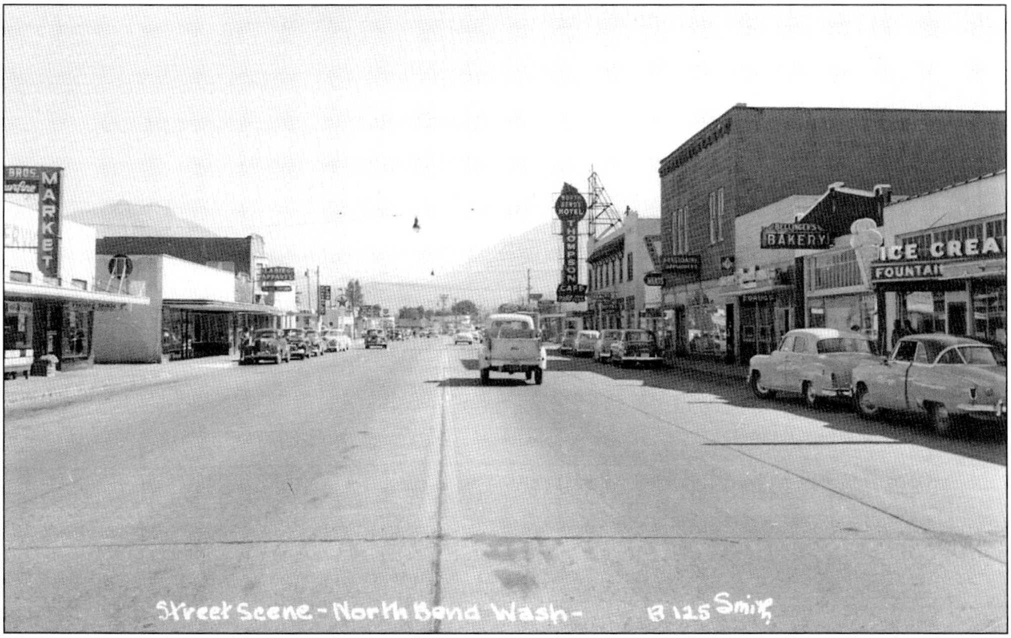

When the highway was widened in 1942, buildings on the north (left, in this photograph) were moved back to accommodate the new four-lane road. Signs for Thompson's Hotel and Café (previously the McGrath Hotel) and Bellinger's Bakery are visible on the right along with signs advertising an ice cream fountain, ladies' apparel, drugs, meats, appliances and autos, and a Texaco station at the far end of town.

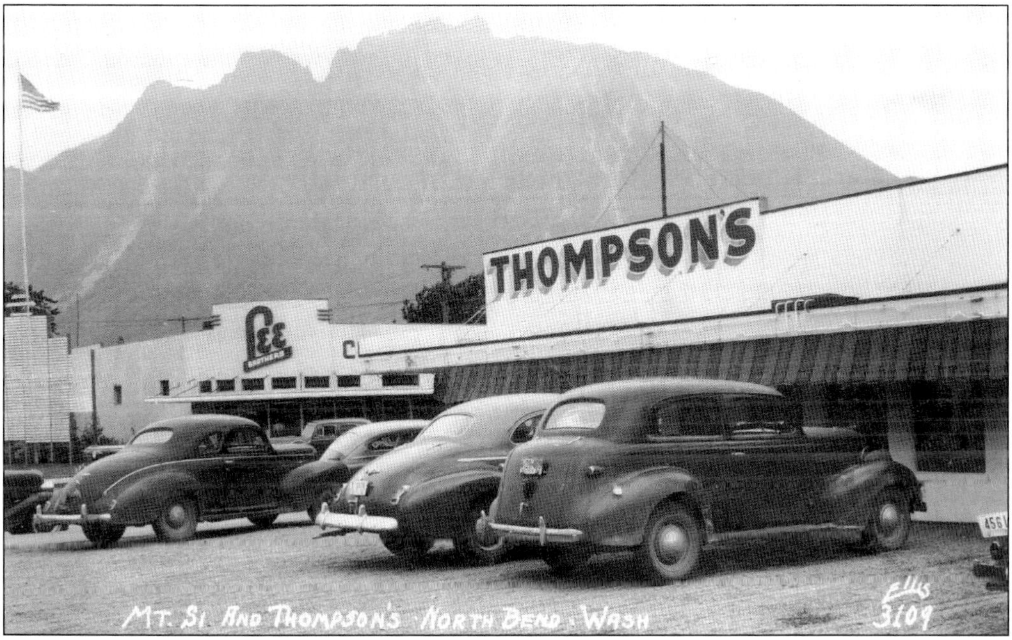

Thompson's Café was originally located at the west end of town. Sometime around 1950 the McGrath Hotel was renamed the Thompson Hotel and the café moved up the street to the same building, whereupon the Mar-T Café occupied the building shown here. It was a featured setting in the television show *Twin Peaks* and today is doing business as Twede's Café.

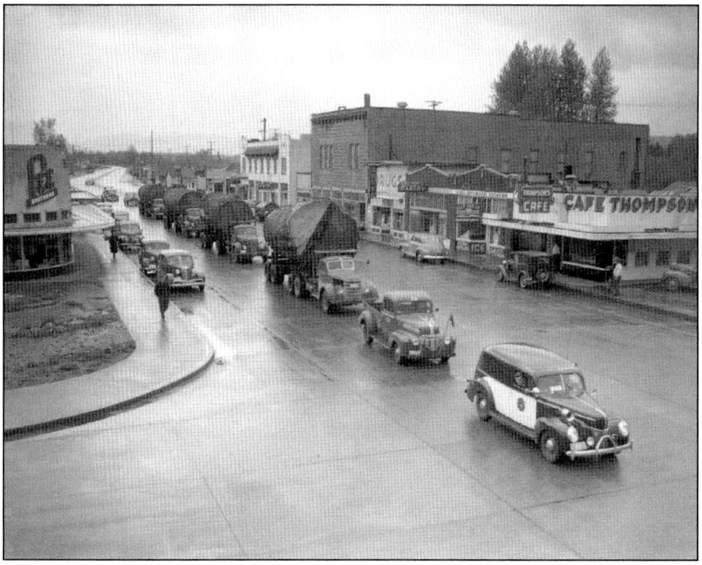

Along with widening the street through town the 1940s road work included construction of a direct alignment of the Sunset Highway between North Bend and Preston. This 1943 photo shows how spacious the new highway was—a 200-foot-tall Douglas fir, cut into sections each hauled by a different truck, easily makes its way through downtown and onto the new westbound route. (MOHAI, 1986.5.5336.1.)

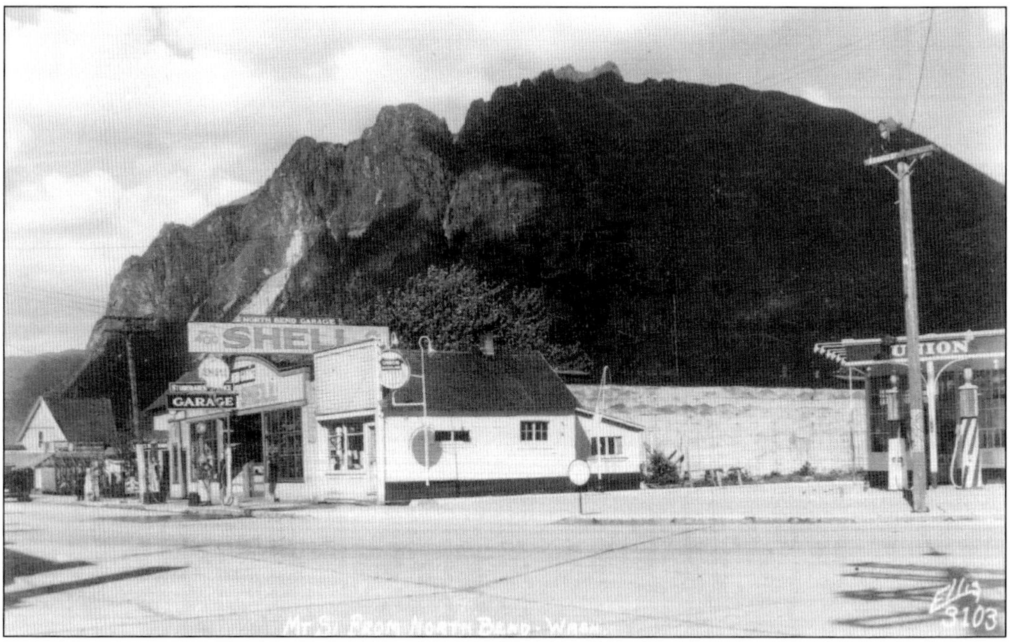

Originally, the Sunset Highway turned north at the west edge of North Bend and followed the Snoqualmie River towards the town of Snoqualmie. These two early service stations—the North Bend Garage and Shell station and the Union station next door—were at the bend of the highway immediately outside North Bend. Mount Si, named after local homesteader Josiah "Uncle Si" Merritt, looms in the background.

Five

North Bend to Seattle

The highway between North Bend and Seattle underwent several revisions—one minor, two major—before its eventual replacement by Interstate 90.

The general route of the Sunset Highway and its predecessor, the Yellowstone Trail, was west from North Bend to Snoqualmie and on to Fall City. At Fall City, the Yellowstone Trail continued west to Kirkland and across Lake Washington by ferry to Seattle, while the Sunset Highway turned south to Preston. From Preston, the Sunset Highway headed west to Issaquah, traveling through downtown to Goode's Corner where it turned south to Renton through the Renton Highlands. Passing through downtown Renton at the southern end of Lake Washington, the highway angled back north to Seattle along Empire Way to a junction with 4th Avenue S (the Pacific Highway) just south of city center.

An early realignment occurred just west of North Bend. In the 1920s, the highway proceeded through town and turned north for a quarter mile before again turning west to cross the Snoqualmie River. Just past the bridge, the original route arced north to Meadowbrook and on into the town of Snoqualmie. The 1930s reconstruction eliminated this course in favor of a more direct route between North Bend and Snoqualmie, bypassing Meadowbrook and the sawmill town of Snoqualmie Falls.

Around 1940, the state of Washington decided to straighten out and shorten the entire route between North Bend and downtown Seattle. The project involved the following three major components: a new highway between North Bend and Preston, bypassing Snoqualmie and Fall City; another highway to pass just outside Issaquah and head west via the Eastgate section of Bellevue to Lake Washington; and the crown jewel of the project a floating bridge to span Lake Washington, with a twin-bore tunnel just west of the bridge to carry traffic into downtown Seattle on Dearborn Street.

When opened end-to-end in July 1940, the project shaved miles and hours off the travel time between Seattle and North Bend. Today it is the route of Interstate 90.

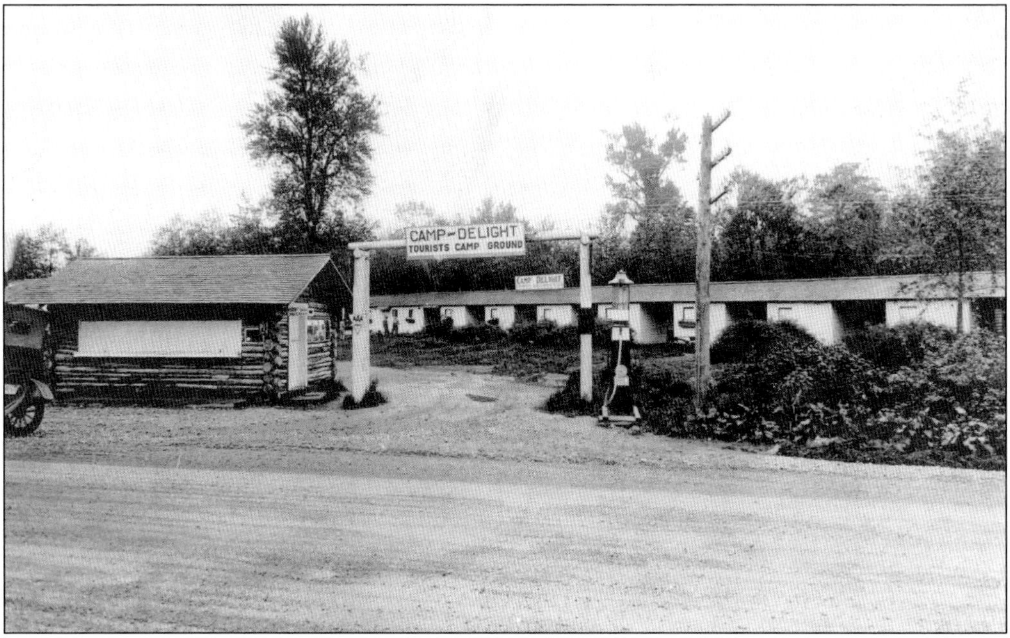

Tourists could find Camp Delight on the outskirts of North Bend, just across the Snoqualmie River bridge west of town. Dating to about 1926, the resort had 22 cottages priced at $1 to $1.50 a night. It was owned by George Gaines, who also ran the local telephone company. Camp Delight was torn down in the 1960s. (SVM, PO.159.0004.)

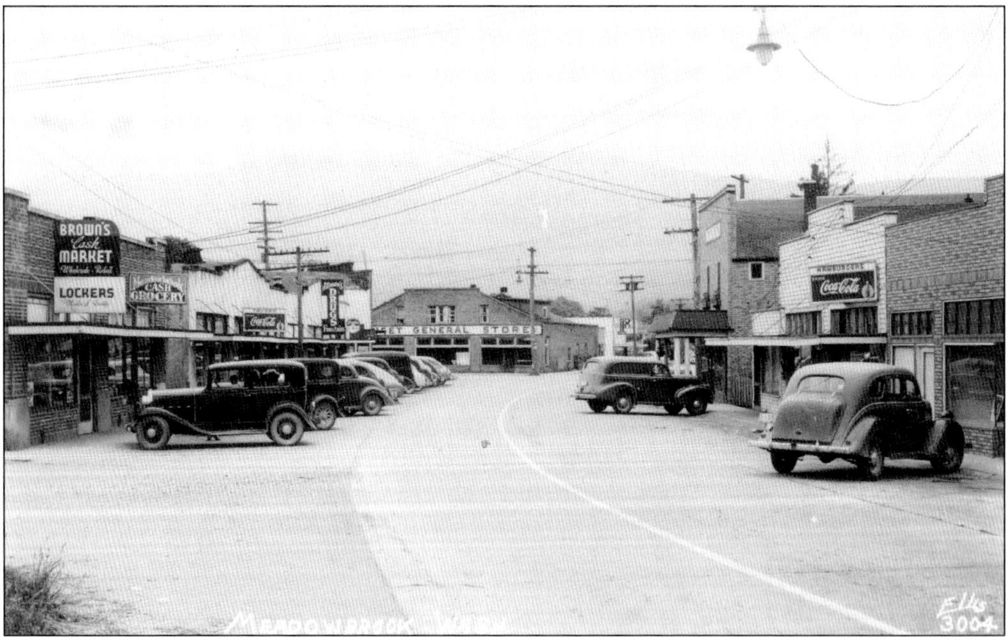

Meadowbrook grew along the original Sunset Highway in the early 1920s. Grocery stores, a hotel, meat market, soda fountain, druggist, and movie theater served travelers as well as residents of the nearby lumber-mill town of Snoqualmie Falls. Highway realignment in 1928 left Meadowbrook off the main route. Most businesses have closed, but it maintains its identity as a separate town, though located just a half mile from Snoqualmie. (SVM, PO.328.4.)

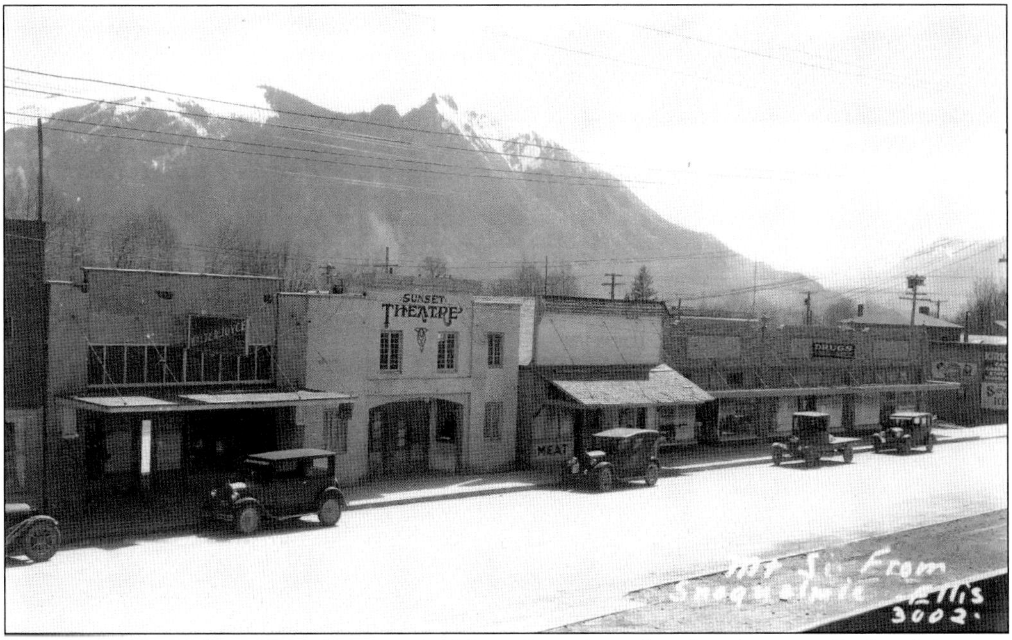

Downtown Snoqualmie's row of business buildings faces the railroad tracks across Railroad Avenue, the main street. The Yellowstone Trail and, later, the Sunset Highway followed Railroad Avenue through town. The Sunset Theatre sits next to Walter Joyce's pool hall in this quiet morning view. Also visible are signs for a meat market, Otto Reinig's store, a druggist, and Kirkman's general store. Mount Si looms in the background. (SVM, PO.014.0067.)

During most seasons, Snoqualmie Falls is a breathtaking sight of thunderous water dropping 268 feet to the rocks below. This is a different view of the falls—frozen in crystalline beauty during one of western Washington's relatively rare cold spells. Overlooking the falls at upper left is the Snoqualmie Falls Lodge; on the opposite bank are two 1900s-era hydropower plants still in use today. (SVM, PO-329-105.)

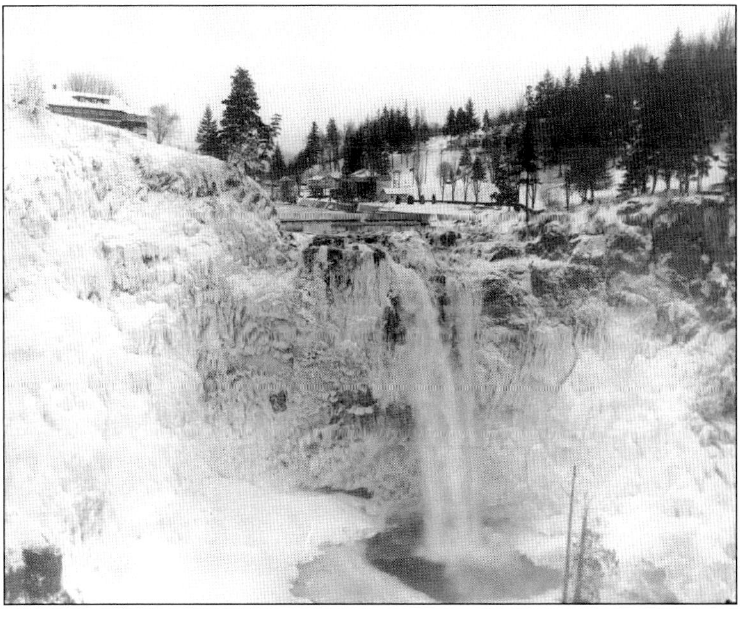

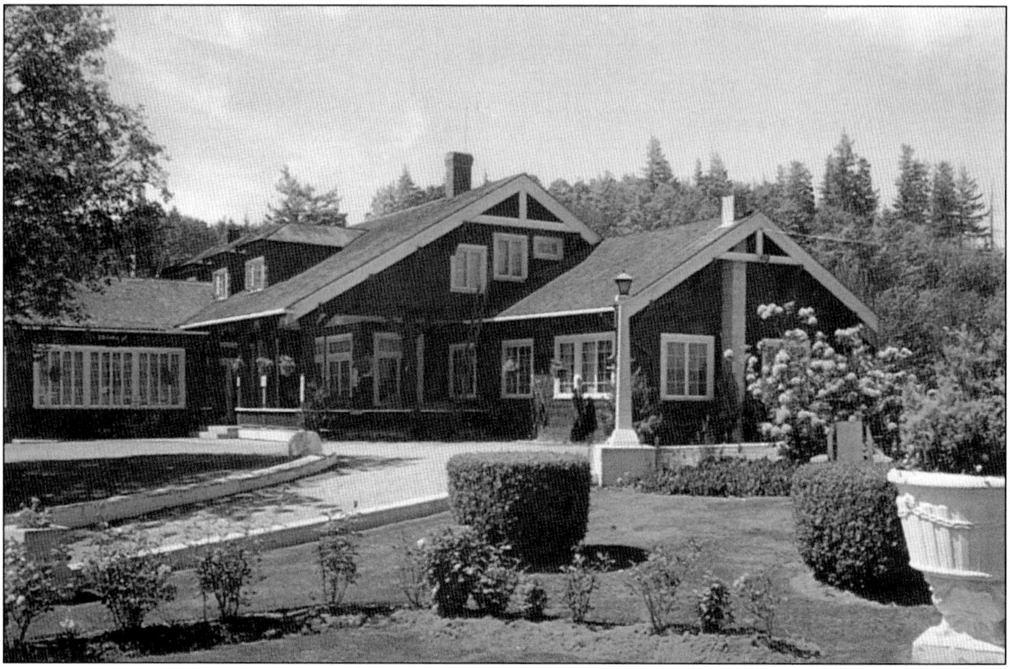

In 1916, an eight-room lodge opened to accommodate tourists visiting Snoqualmie Falls. Perched on the edge of the falls, the Snoqualmie Falls Lodge has greatly expanded over the past hundred years—completely remodeled in 1988, it was rechristened the Salish Lodge—and is now a world-class retreat. The lodge had a starring role in the 1990s television show *Twin Peaks* as the Great Northern Hotel. (JCO'D.)

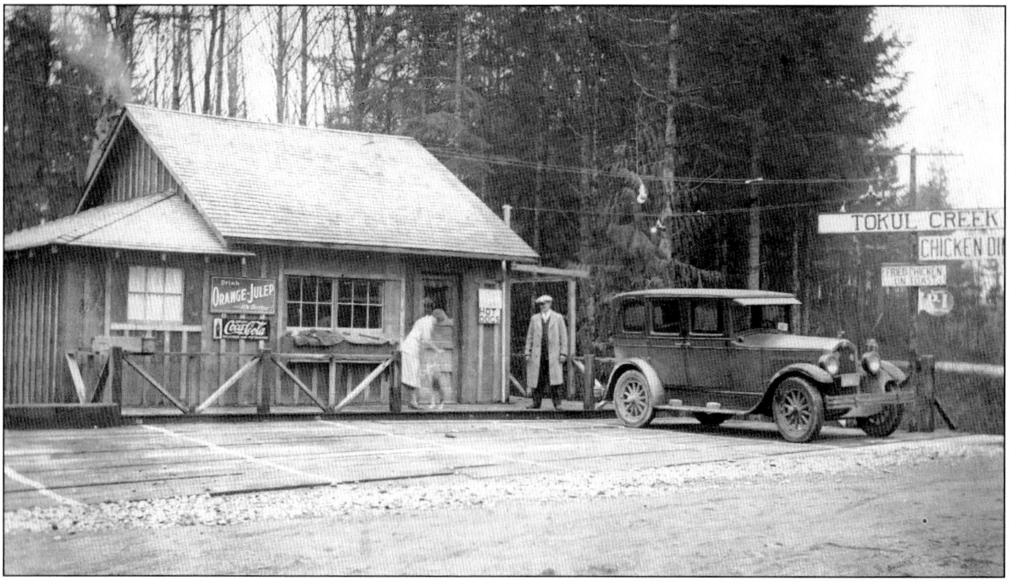

Just west of Snoqualmie Falls, the Sunset Highway descended 300 feet to river level, crossing over Tokul Creek on the way down. This rustic inn, built partly over the creek, sat alongside the highway at a hairpin turn on the way down. Patrons were offered chicken dinners and fried chicken on toast as well as hot dogs and chili; drink choices included Orange Julep and Coca-Cola. (SVM, PO.124.0088.)

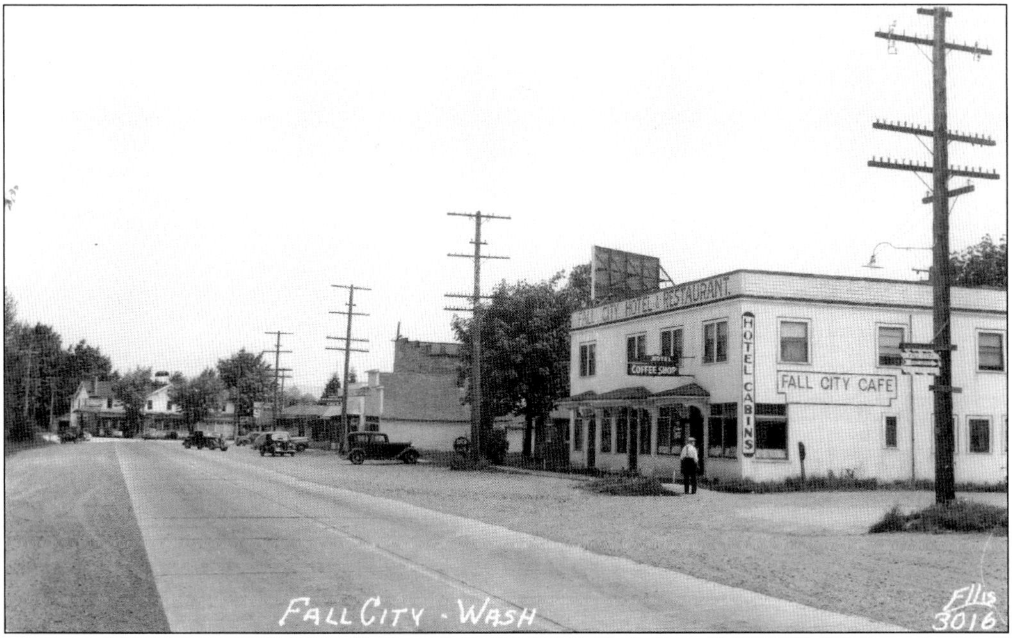

In early days, Fall City was known as "The Landing." With Snoqualmie Falls blocking river travel upstream, it was the unloading point for eastbound freight. The Yellowstone Trail and the Sunset Highway split at Fall City, with the Sunset Highway heading south along the Raging River to Preston and the Yellowstone Trail continuing west to Kirkland and, by ferry, into Seattle. (SVM, PO.430.7.)

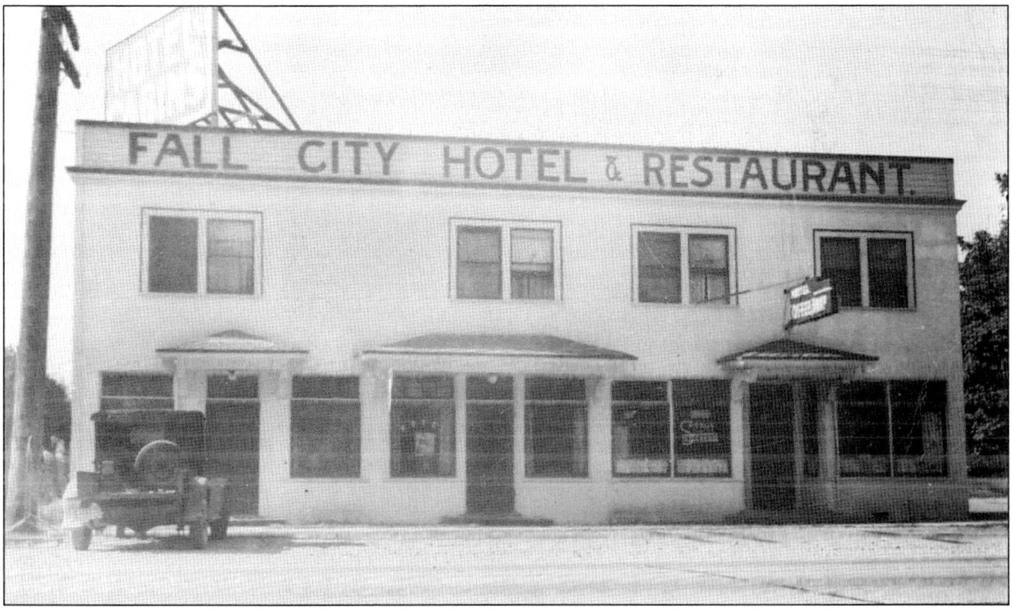

The Fall City Hotel was in business as early as the turn of the 20th century. Offering both hotel accommodations and cabins, it was a landmark for early motorists, situated as it was at the junction of two important highways. The hotel burned in a spectacular fire in September 1925 that nearly destroyed the whole town. Frank Thomas rebuilt and operated the hotel until his retirement in 1948. (SVM, PO.30.82.)

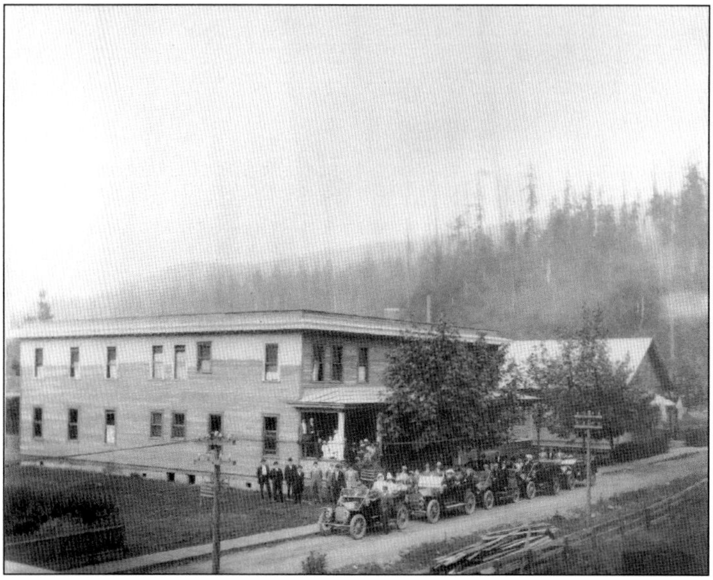

The Hotel Preston, built in 1912, originally served business visitors to the small lumber mill town of Preston. When the Sunset Highway was routed through town, the hotel repurposed itself to accommodate motorists. Twenty-six rooms were available, each with steam heat, running water, and private bath; the grounds included a tennis court with good fishing nearby in the Raging River. The hotel was demolished in 1977. (SVM, PO.219.31.)

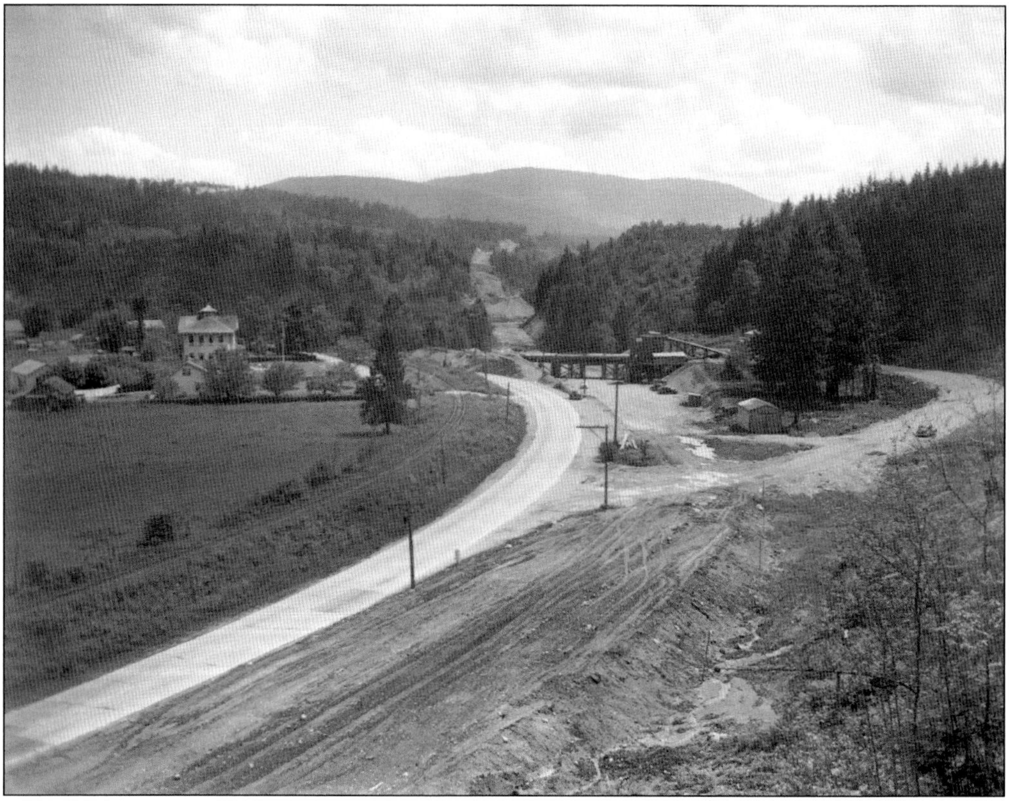

Highway realignment in the 1940s straightened the route between North Bend and Issaquah. Heading directly west from North Bend to a junction with the older highway at Preston, it completely bypassed Snoqualmie and Fall City. In this view looking east, the original highway, having climbed along the Raging River from Fall City, bends towards Issaquah; the new construction proceeding from North Bend is seen beyond the curve. (MOHAI, PI25313.)

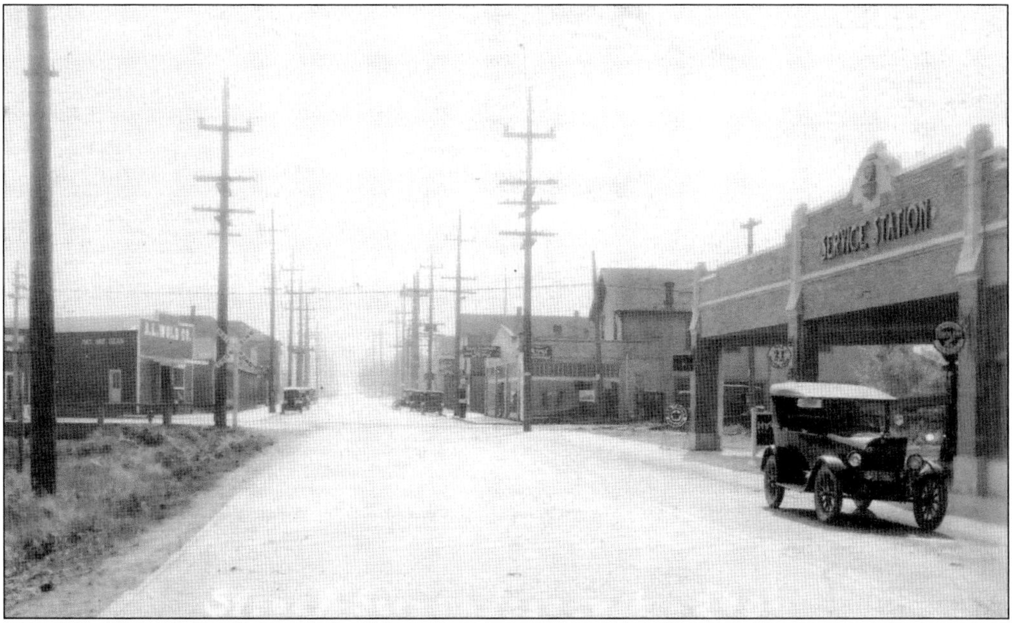

This c. 1925 view looks east along Sunset Boulevard to the intersection with Front Street in downtown Issaquah. On the right, a service station advertises Red Crown gasoline and Zerolene motor oil; portions of the building are still in use as the Issaquah Brewhouse. In the background, A.H. Wold's livery stable squares off against Case & Hepler's Ford dealership, its nemesis across the street. (UW, WAS1084.)

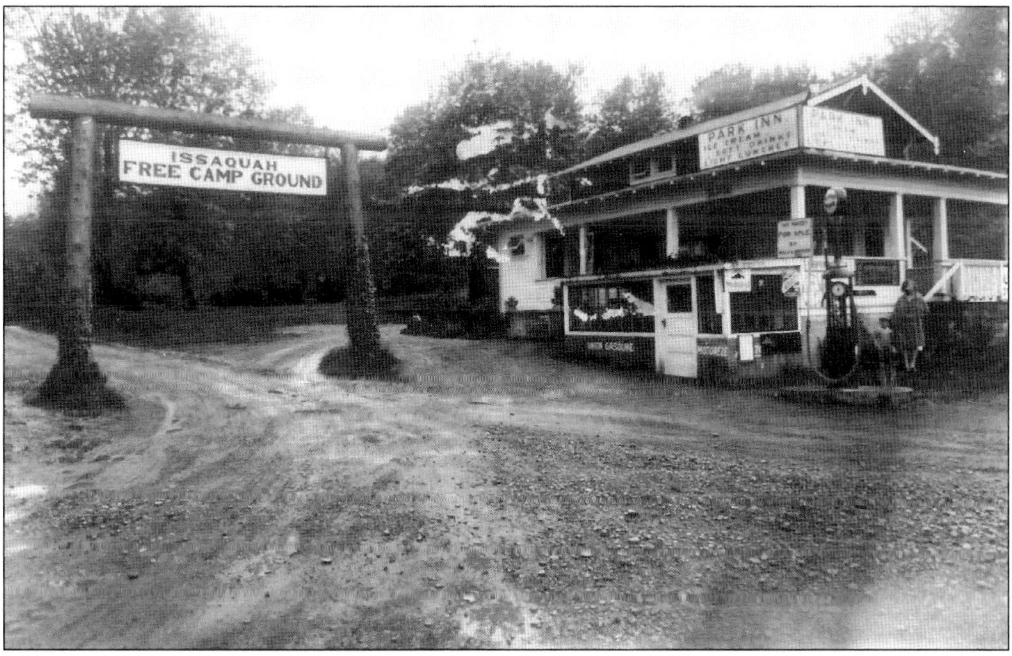

A "For Sale" sign is tacked to the Park Inn in this c. 1927 photograph. The inn adjoined the Free Camp Ground, which Issaquah had opened in 1921 in the city park. Purchased by Ed and Claire Leslie in the early 1930s, the inn was acquired in 1937 by William Kiebert, who operated it until his death 12 years later. Apartments now occupy the site. (IHS, 95.009.001.)

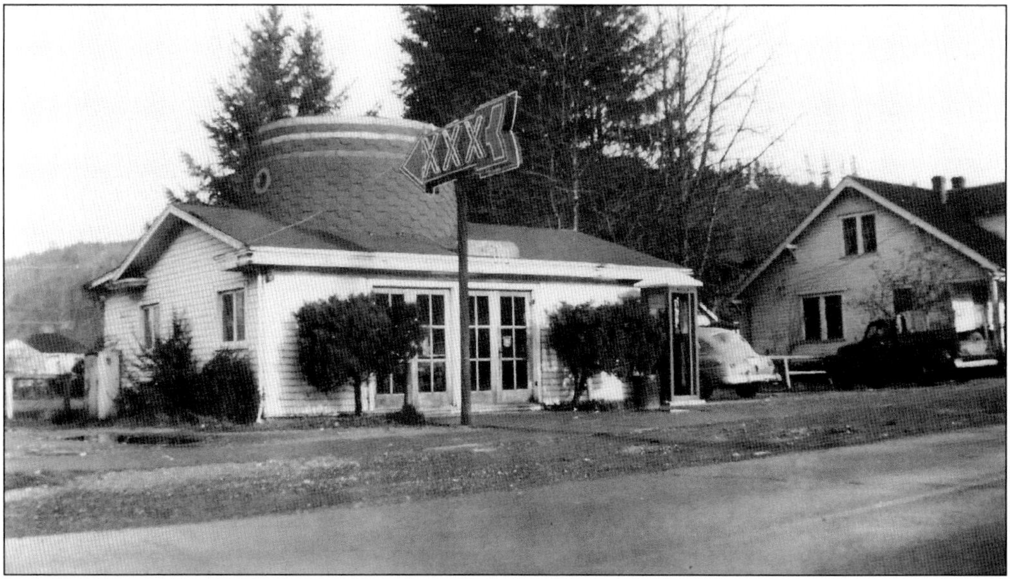

This Triple XXX restaurant stood in a residential section along East Mill Street (now Sunset Way) for many years. Owned by Dave Morgan in the 1940s, it was eventually demolished to make way for an auto dealership. The barrel—unusual in being covered by shingles rather than planks, as were virtually all other Triple XXX barrels—towers over the house onto which it was grafted. (IHS, 2008.006.003.)

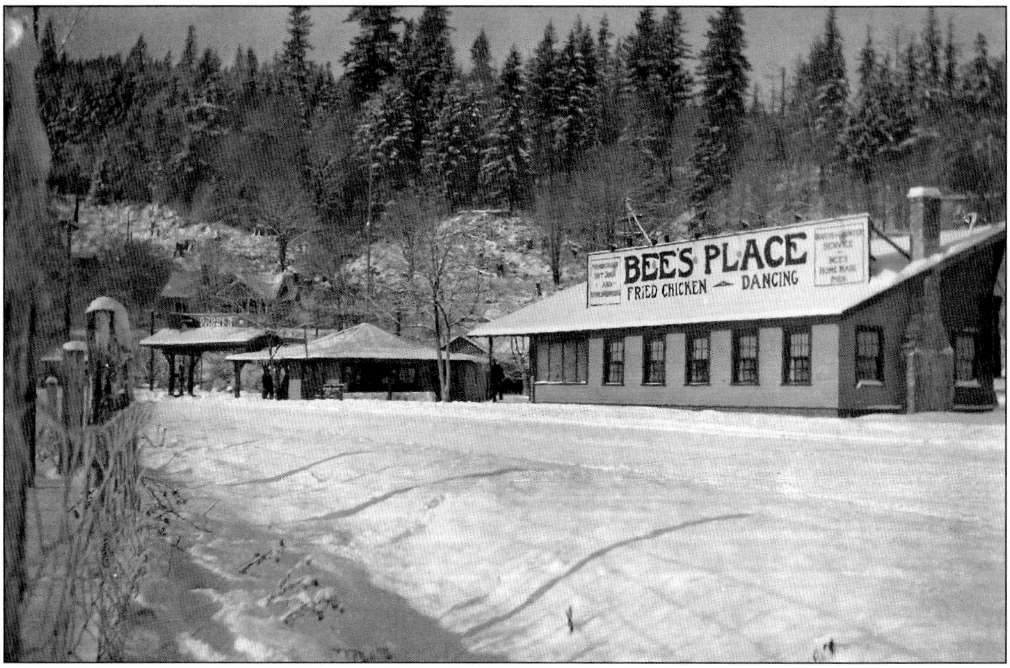

Bee's Place celebrated its May 19, 1928, grand opening with a chicken dinner and dancing. "Bee" was Cora Bea (Goode) Lassen; the location was Goode's Corner, two miles northwest of Issaquah. In 1934, L.J. Livingston purchased the place and reopened it as Livingston's Sammamish Inn, serving sandwiches and beer. A shooting affair in January 1941 sent three men to the hospital—proprietor Ernest Hale was arrested but cleared. (IHS, 2005.004.013.)

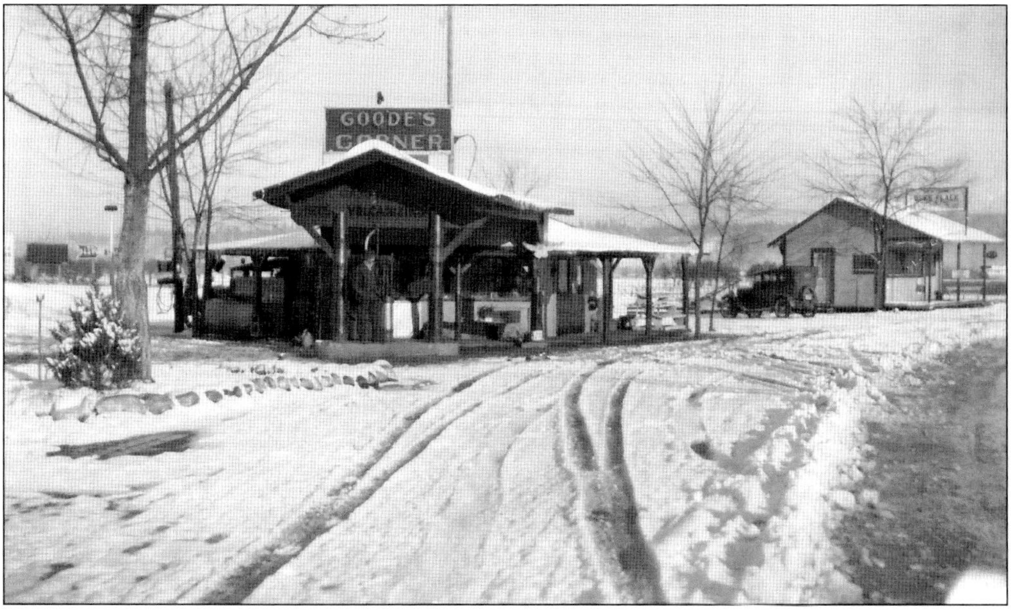

Motorists heading west were well advised to check their fuel gauges at Goode's Corner: just past Bee's Place the Sunset Highway turned south, passing through nine miles of sparsely populated hill country on its way to Renton. Luckily, the Goode's Corner service station stood by to ease the minds of those who feared they might run low on fuel before reaching the next services. (IHS, 2005.004.014.)

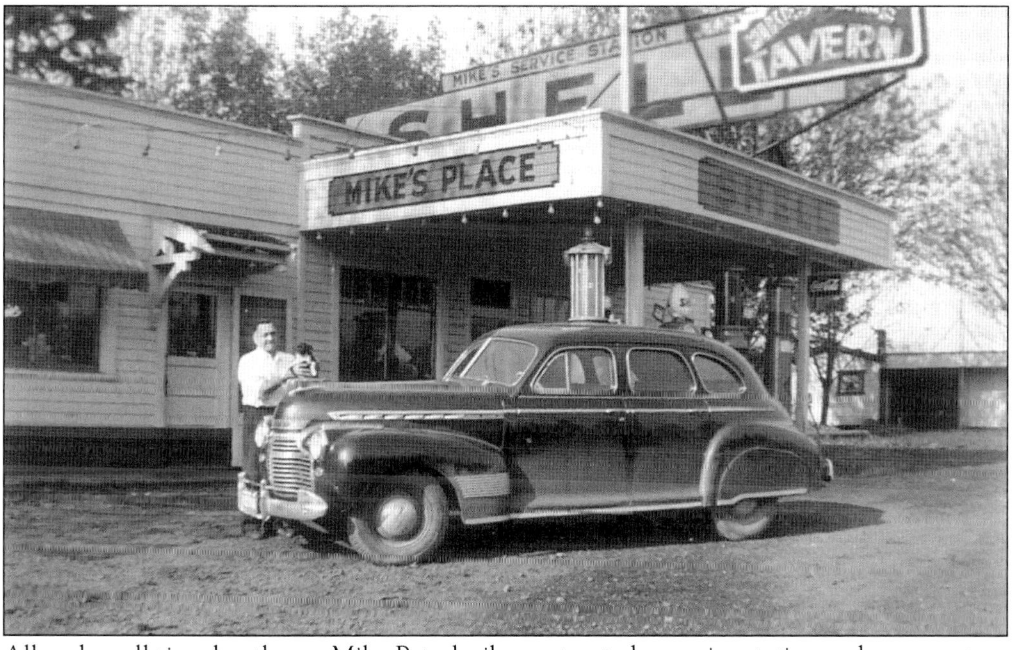

Alleged small-time bootlegger Mike Potoshnik constructed a service station and grocery store on the Sunset Highway east of Renton prior to 1930, choosing a location outside the city's legal jurisdiction. An auto camp and cabins were soon added. By 1940, the cabins had disappeared and a tavern—Mike's Place—had opened in the remodeled gas station. The building is now gone. (RHM, 1993.144.3664.)

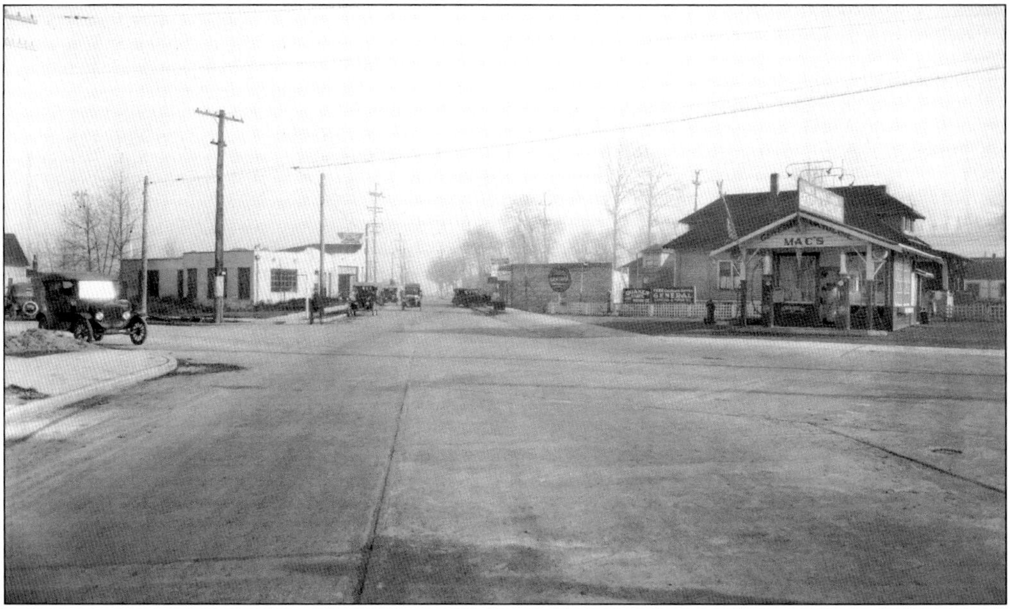

Robert "Mac" McDonald's Service Station was one of several auto-oriented businesses Sunset Highway motorists encountered just past the Cedar River bridge into Renton. Another garage was across the street, and up the block was Clarke Brothers' Ford dealership. Mac sold General gasoline and Quaker State motor oil at his two-pump station, now the site of an insurance office at Bronson Way and Main Street. (MOHAI, 1983.10.3068.2.)

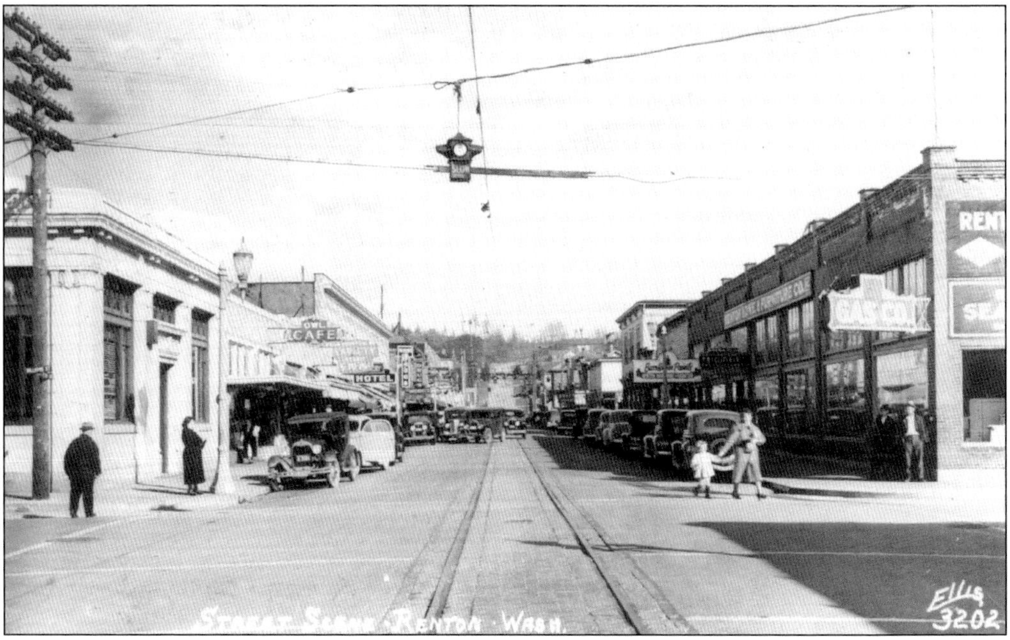

It was busy in downtown Renton the day this photograph was taken, with cars jockeying for position and pedestrians lining the sidewalks. Streetcar tracks are still in place, soon to give way to autos. The view looks east from Williams Avenue along 3rd Street, the Sunset Highway route through the city. Signs include the Owl Café, Renton Variety Store, and Renton Home & Furnishing Co.

Frank Rhodes opened a filling station in downtown Renton in 1924. By 1940, he had expanded his business to include Firestone Tires, as advertised by the sign hanging off the building that housed the tire and repair business. In later years, he operated a Studebaker-Nash-Packard dealership a block west of his original location near the Renton Theatre. (MOHAI, 1983.10.16648.)

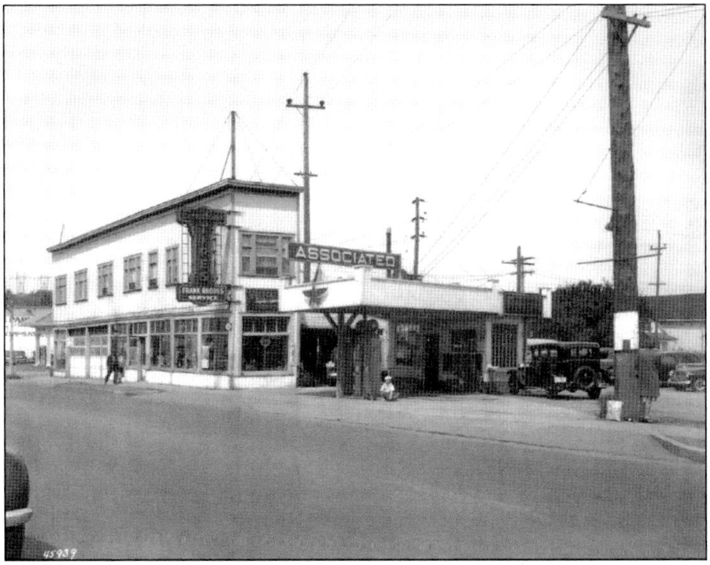

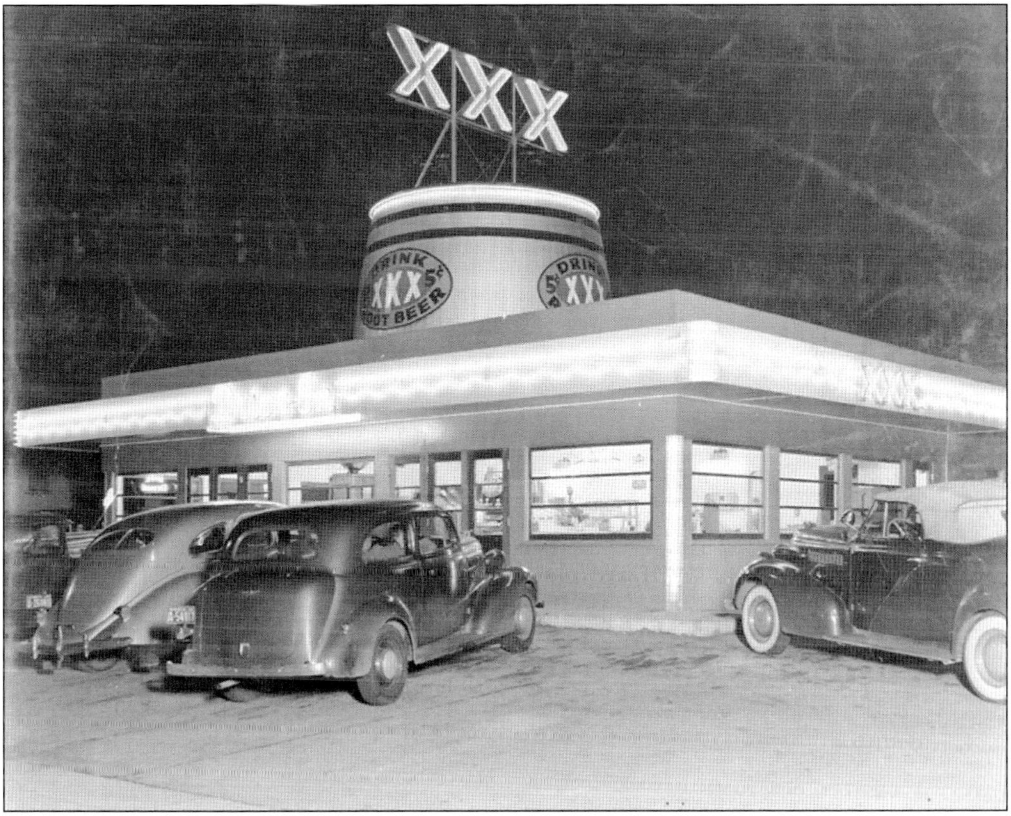

Montana-born H.J. Rutherford opened his first Triple XXX barrel in 1932 in Renton. By 1940, he and sons A.H. and L.W. were operating 10 Triple XXX franchises in Seattle and Spokane as well as the original in Renton shown here, all featuring this distinctive design of a giant barrel atop a modernistic, neon-fringed restaurant providing both curb service and indoor seating. (RHM, 2001.130.5658.)

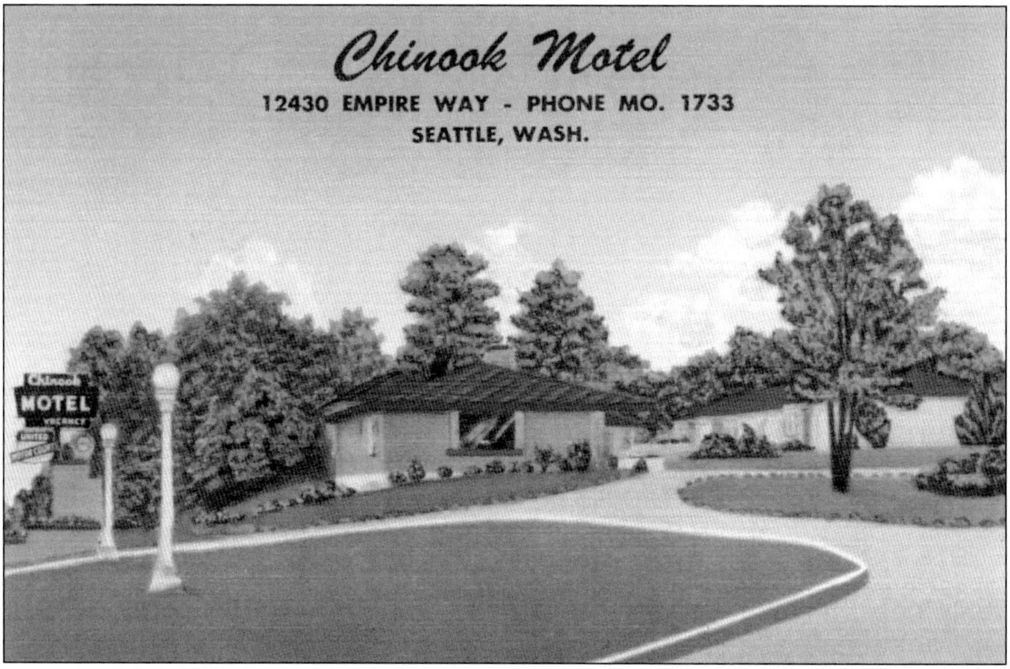

Roadside lodgings were scarce between Renton and Seattle. One of the few places to be found was the Chinook Motel at 12430 Empire Way (now Martin Luther King Jr. Boulevard), three miles north of Renton. The 12-unit auto court was owned by Mr. & Mrs. Barrington Upper and offered "hotel quality" rooms with tiled baths and electric kitchens. Empty and deserted, the Chinook was still standing in 2013.

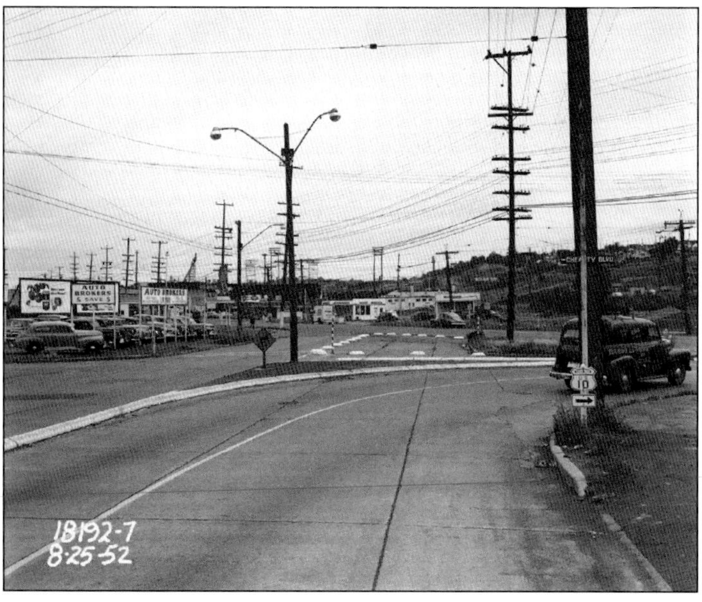

Empire Way intersected another major arterial road, Rainier Avenue, about two miles southeast of Seattle, then turned towards the city. Although Highway 10 was officially crossing Lake Washington by the time this photograph was taken in August 1952, a highway shield displaying the US 10 logo is clearly evident. Kuehnoel's Triple XXX Barrel can be seen opposite the Richfield eagle, one of two gas stations occupying the intersection. (SMA, 43940.)

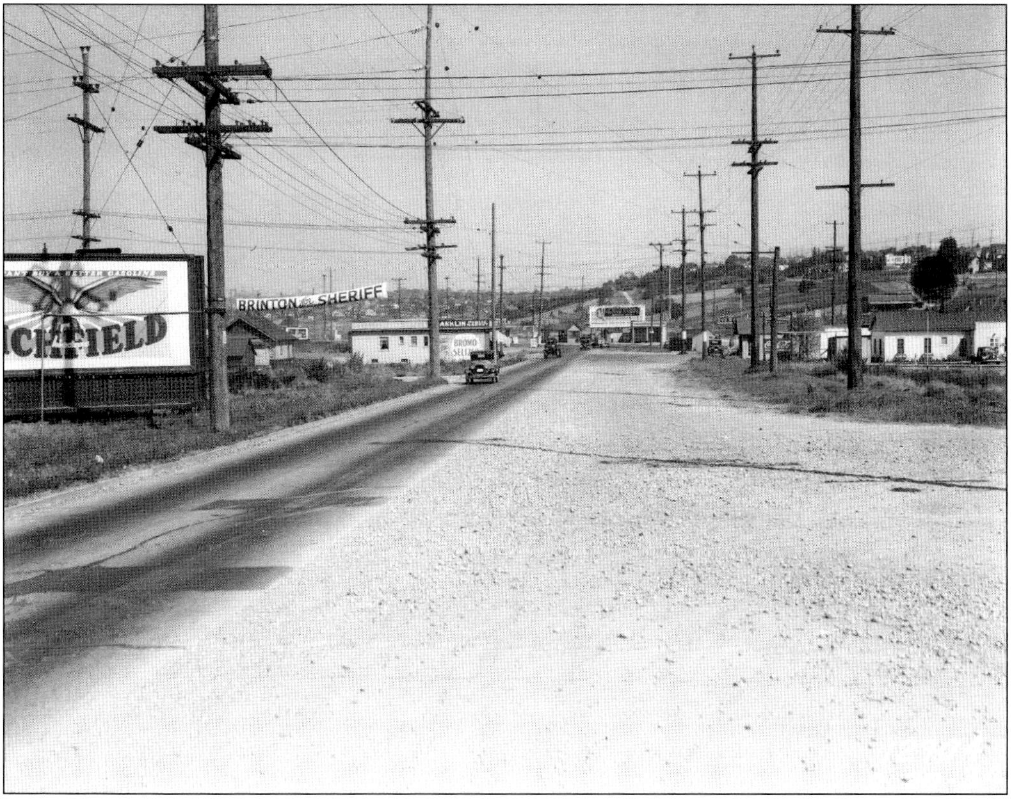

This August 1934 photograph shows a recently paved Empire Way. Heavily built-up today, south Seattle was only lightly developed at the time. Several service stations—Shell and Richfield—line the highway along with billboards advertising Bromo-Seltzer and the Exeter Hotel in downtown Seattle and a banner urging citizens to vote for Brinton for sheriff. (SMA, 9081.)

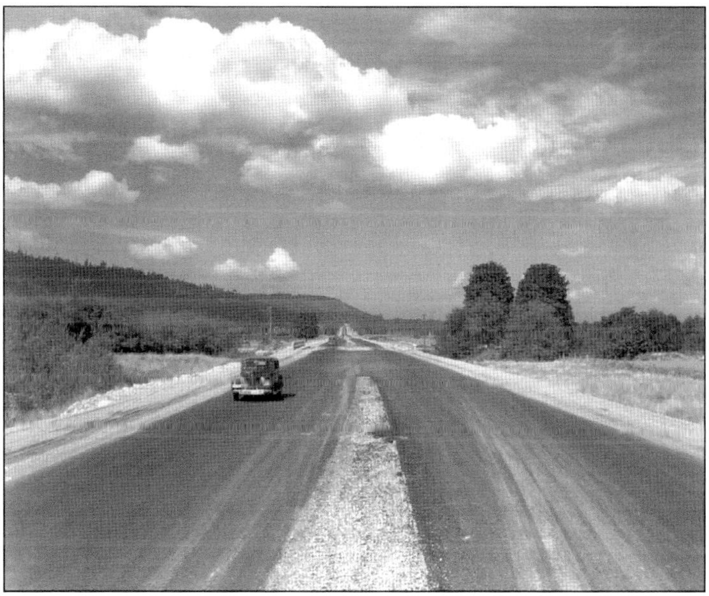

The 1940s Sunset Highway realignment between North Bend and Seattle curved northwest at the edge of Issaquah, bypassing the business center. It then headed west to the recently completed Lake Washington Floating Bridge, cutting off the loop around the south end of the lake through Renton. In this August 1940 photograph, the highway just west of Issaquah is so new it has not yet received centerline stripes. (MOHAI, PI25307.)

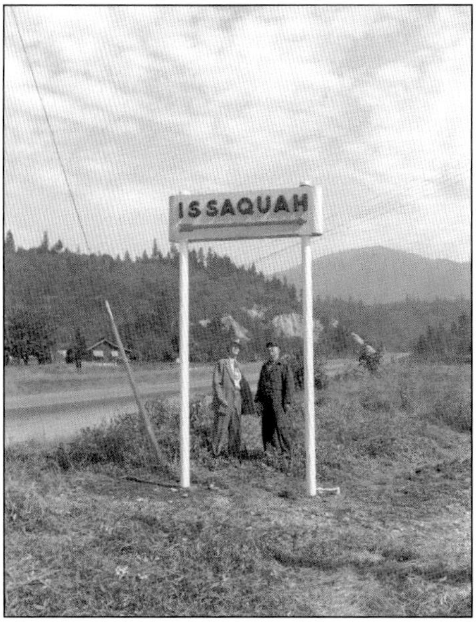

Concerned that the new bypass would deflect business, Issaquah residents waged an 18-month campaign with the highway department to erect a sign directing highway travelers into downtown. By August 1948, a tall lighted sign with green letters and an amber arrow—seen being admired by Earl Bramble of the Issaquah Chamber of Commerce and city council member Alfred Keogh—pointed motorists in the right direction. (MOHAI, 1986.5.4492.)

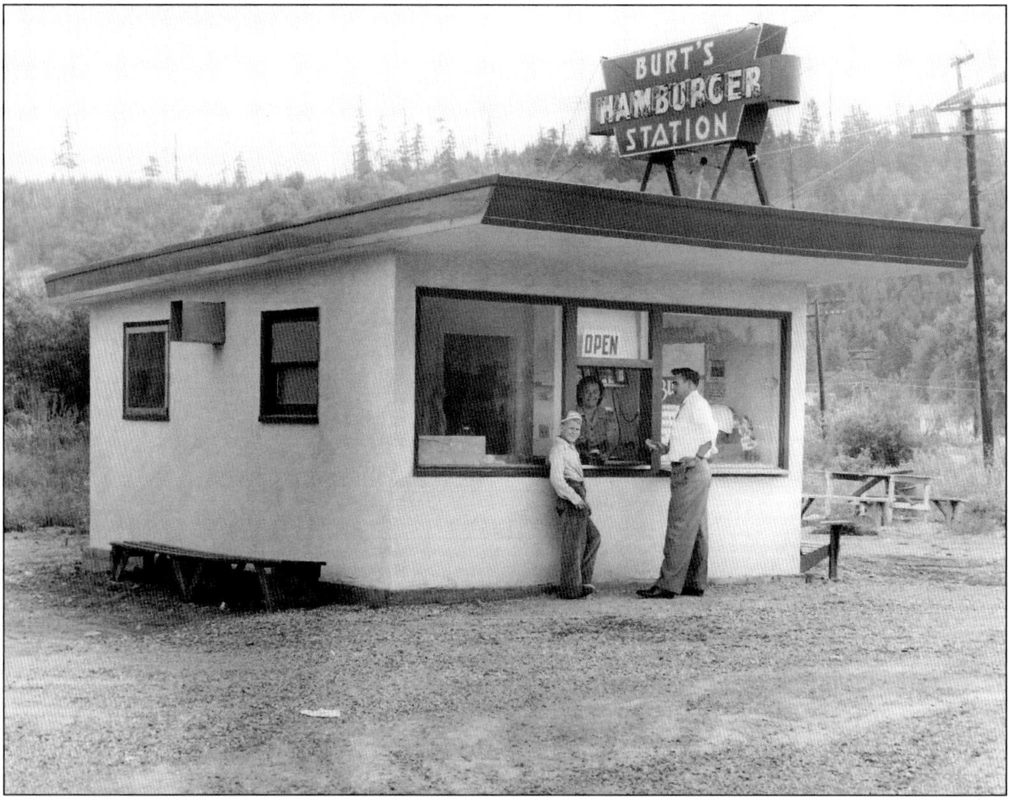

Issaquah need not have feared a loss of business from being bypassed. Travel through downtown was impacted, but it was not long before the new highway attracted businesses catering to travelers. Burt's Hamburger Station was one such place. Owned by Clarence and Mabel Burt, it was a prototypical 1950s drive-in along the highway just west of town. Like many early roadside attractions, Burt's has disappeared beneath Issaquah's explosive growth. (IHS, 72.021.014.071.)

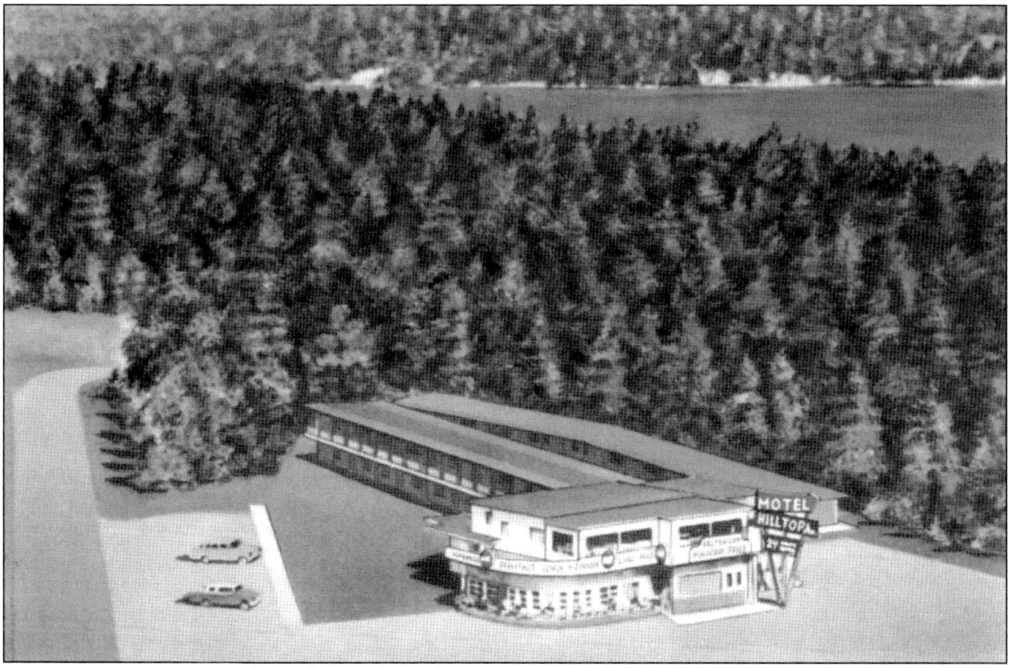

The Hill-Top Inn Motel, located along the new highway atop the ridge separating Lakes Sammamish and Washington, commanded a panoramic view of the lakes and the Cascade Mountains. With 36 units, all with radiant heat, television, and telephones, the Hill-Top was billed as the Eastside's most luxurious and modern highway hotel. A coffee shop and restaurant were adjacent. Rustic at the time, the site is heavily developed today.

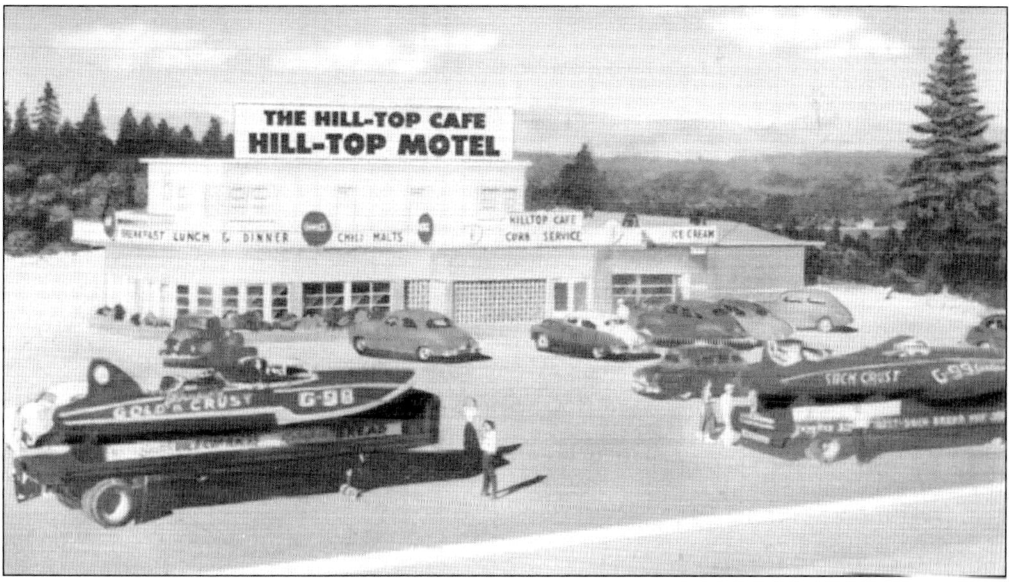

The Hill-Top Cafe, the restaurant at the Hill-Top Inn, was owned by Billy Lind. Open from 6:00 a.m. to 2:00 a.m. daily for dinner and dancing, its menu featured steaks, chicken, and seafood along with fountain service and short-order items such as chili. In this view, patrons are admiring two hydroplanes, both sponsored by bakeries (Gold'N Crust and Such Crust Bread), bound for the annual hydro races on Lake Washington.

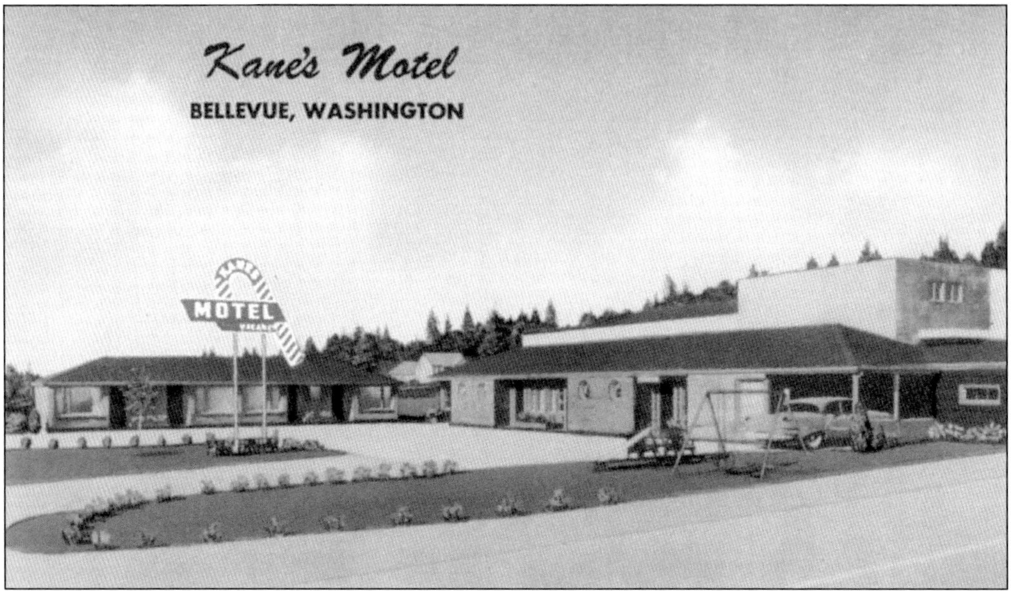

In keeping with their name, Jim and Maxine Kane advertised their Kane's Motel with a distinctive candy-cane sign. The 17-unit motel stood at 14644 Sunset Highway in the Eastgate section of Bellevue, a rapidly-developing area along the Sunset Highway at its intersection with 148th Avenue SE. Kane's is no longer in business, but the building and sign still stand under the shadows of Interstate 90 access ramps.

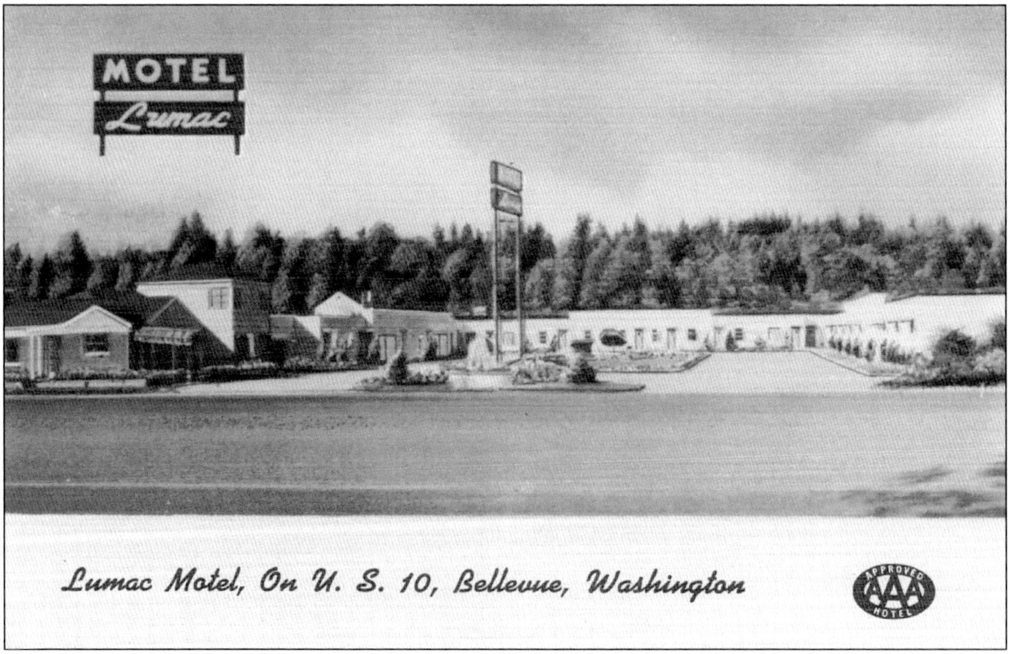

Owners H.E. and Luttie McKinley combined the first letters of Luttie's names to come up with Lumac. It sat right next door to Kane's at 14632 Sunset Highway. The Lumac expanded over time, growing from 22 to 28 "new and fully modern" units, all with kitchens, television, and radios; in 1954, the nightly rate was $5 to $7. Development has overwhelmed Lumac's, and the Silver Cloud Hotel occupies the site today.

In 1921, engineer Homer M. Hadley proposed a bridge to cross Lake Washington using concrete pontoons—a new technology at the time. Hadley argued pontoons were the only feasible technique, given the lake's depth and soft bottom. It was not until 1937 that his ideas found acceptance. Construction of the $9 million, 6,620-foot bridge started in 1939. This view of the bridge under construction looks west to Seattle. (MOHAI, PI20571.)

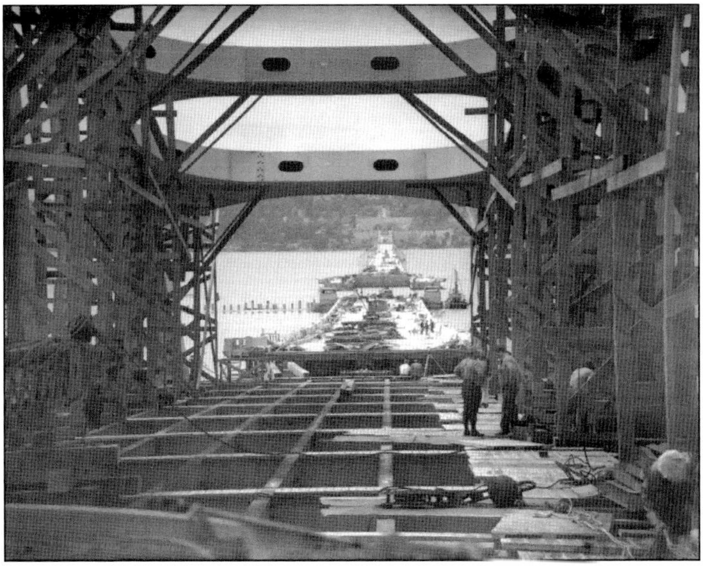

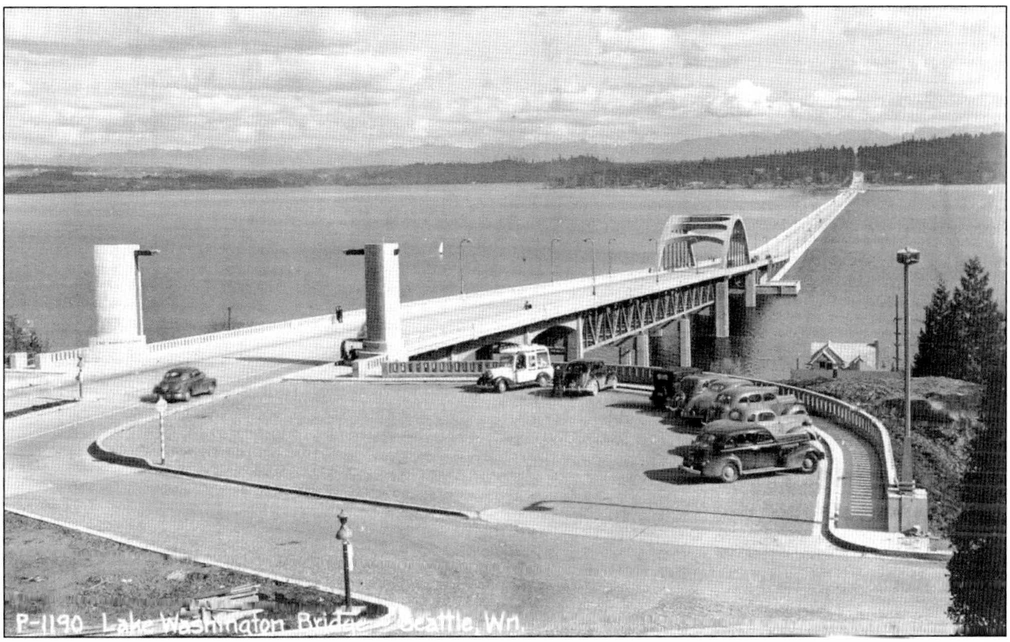

The bridge, officially named the Lacey V. Murrow Memorial Bridge, opened on July 2, 1940, with appropriate ceremonies: 2,000 people attended, politicians spoke, bands played, and ribbons were cut. An immediate success—11,611 cars crossed the span the first day, though average daily figures were projected to be 2,800 cars—the bridge shaved an hour of travel off the previous route via Renton.

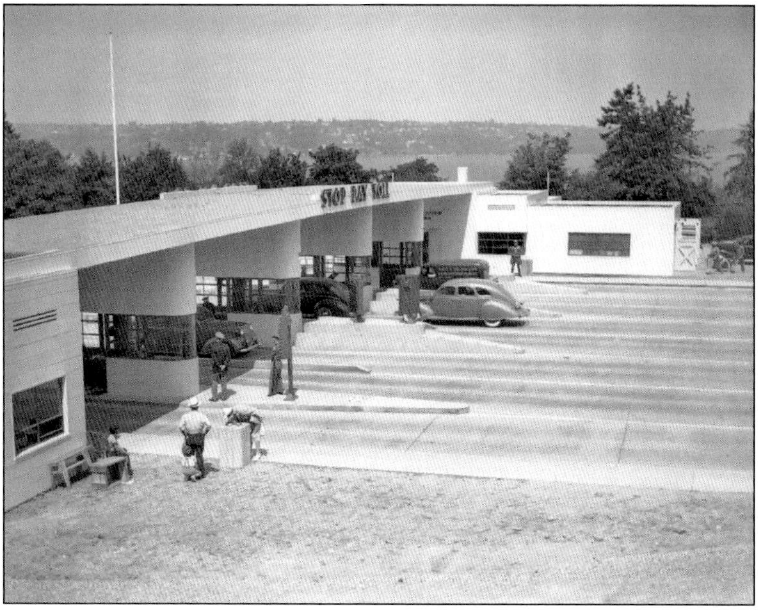

After the opening ceremony, Gov. Clarence D. Martin was first to pay to cross the bridge. Tolls were 25¢ for auto and driver, 5¢ each passenger, and 5¢ for pedestrians. The toll schedule even included a charge for horse-drawn wagons—35¢ to 50¢ depending on how many horses. When tolls ended in 1949, the celebration was as big as the bridge dedication. (MOHAI, PI20590.)

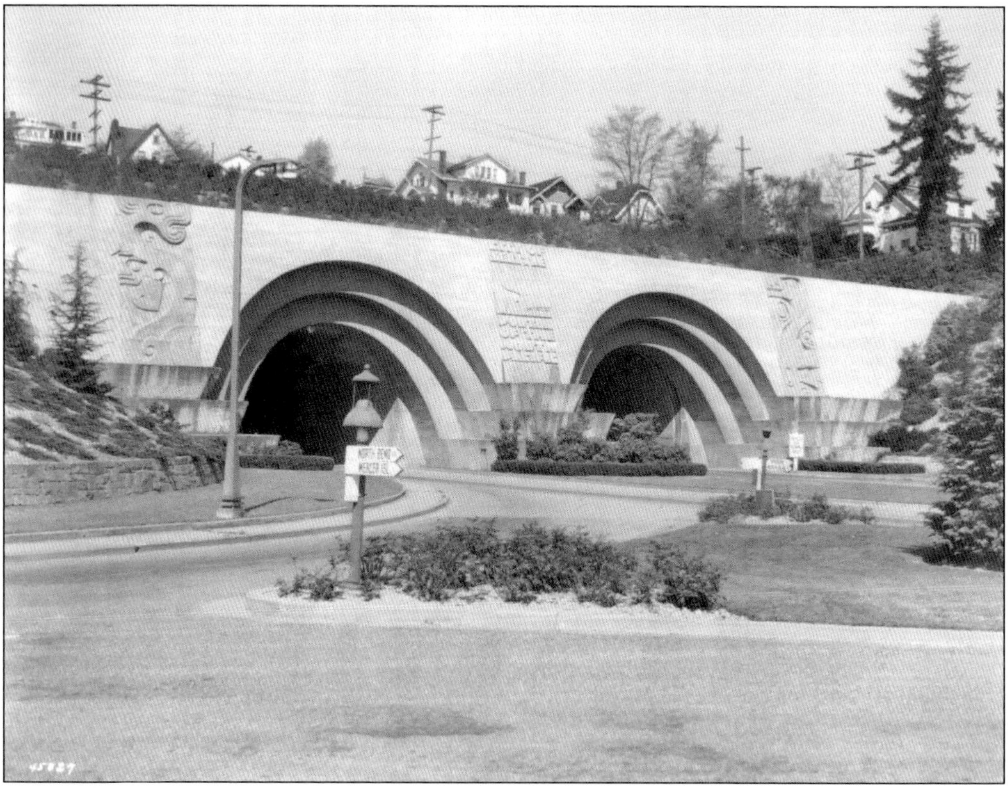

A tunnel was required under Mount Baker, the ridge between the west shore of Lake Washington and downtown Seattle. The 1,466-foot-long, twin-bore tunnel was opened simultaneously with the floating bridge. Its east portal, with its stylized Native American motifs, welcomed motorists to Seattle, "the Portal of the North Pacific." These two bores, integrated into Interstate 90, now carry eastbound traffic; a new adjacent tunnel handles westbound flow. (MOHAI, 1983.10.15525.)

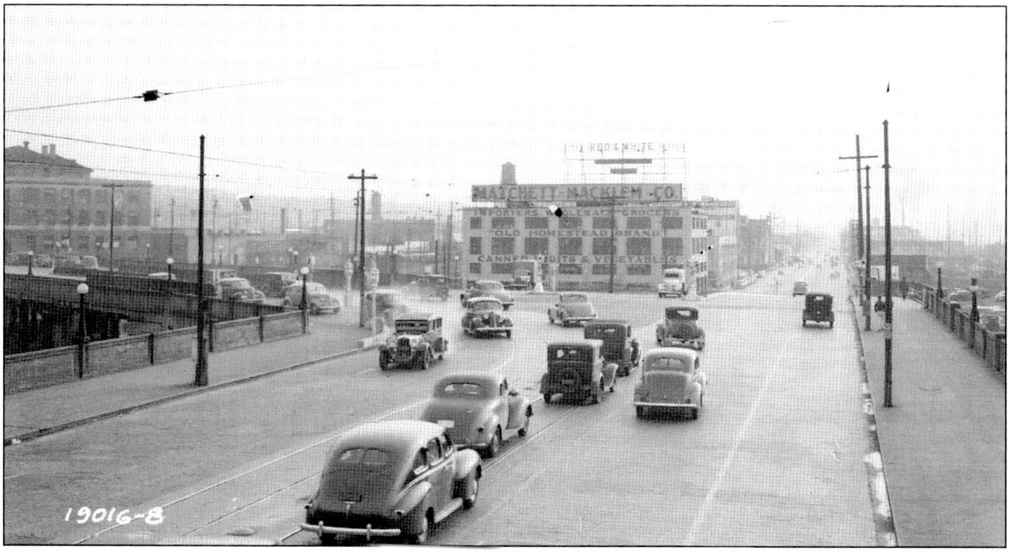

Washington's great east-west and north-south highways meet in Seattle: The Sunset Highway (Highway 10) and the Pacific Highway (Highway 99) join at the junction of Airport Way and 4th Avenue S in this view looking from the north. The intersection was just as busy then as today. This photograph dates to 1950, although many of the automobiles appear to be from an earlier era. (SMA, 54606.)

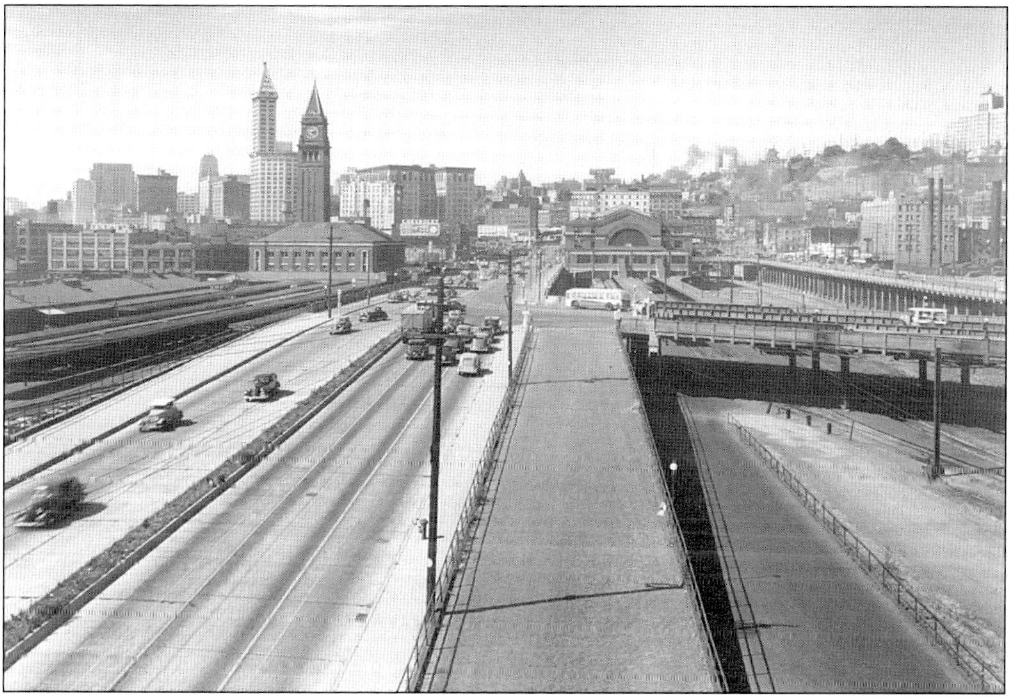

The end of the line, Highway 10 entered Seattle northbound on 4th Avenue S in this 1945 view. Visible here is the Smith Tower, completed in 1914; until 1962, it was the tallest building west of the Mississippi River. King Street Station (serving the Great Northern and Northern Pacific Railways) is to its east; across the street is Union Station. The Frye Hotel is just north of King Street Station. (MOHAI, SHS6141.)